CAKE DESIGN AND DECORATION

Printed by The Continental Printing Company
Limited, Hong Kong.

CAKE DESIGN AND DECORATION

by

Bernice Vercoe and Dorothy Evans

MURRAY

SYDNEY :: MELBOURNE

Published by The K. G. Murray Publishing Company Pty. Ltd.,
142 Clarence Street, Sydney, N.S.W.

First Impression, 1966
Second Impression, 1966
Third Impression, 1967
Fourth Impression, 1968
Fifth Impression, 1970

Contents

Some of the many Show Trophies and Ribbons awarded to the authors.

Foreword

MAY WE WELCOME you to the wonderful world of cake decorating, hoping that you may derive as much pleasure in reproducing the moulded work and cakes as we did when composing this book.

We have endeavoured to introduce new trends into our designing and feel sure that if the instructions are carried out, the finished product will be very pleasing to the eye.

In cake decoration there is unlimited potential for those with imagination and we hope that, coupled with your own ability, we may be a help with some new ideas. For the person lacking that flair, we hope this book will be of even greater benefit.

Many cakes illustrated in this book have been prize-winning exhibits from various shows and all have been done by ourselves, with the exception of three novelties, namely the pot of roses, the vase of sweet peas and the novelty table.

These were prepared by Mrs. D. Scott, a 75-year-old decorator, showing that the art of cake decorating is not reserved for the younger generation. Opposite is a display of the many ribbons and trophies won by us for our entries with show cakes.

B. J. VERCOE and
D. EVANS.

Introduction

THIS BOOK IS intended as basic instruction for the beginner in conjunction with a cake decorating course and as a refresher for the advanced student. It touches on the principles of general piping and cake ornamenting.

If the student is prepared to do plenty of practice in conjunction with the information supplied by this book, results of high standard may be achieved.

The main essentials of materials and equipment for decorating and piping are few. They are:

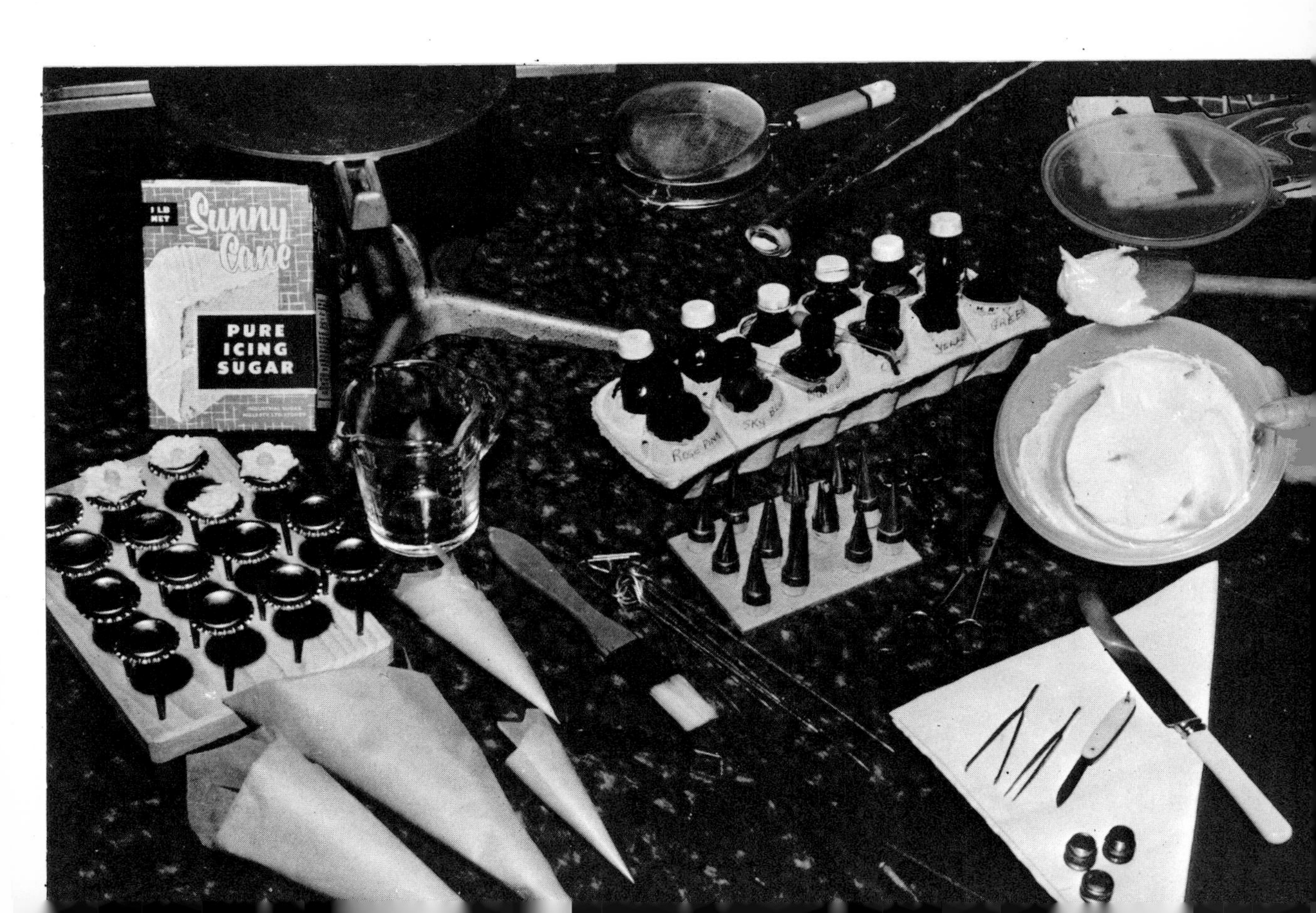

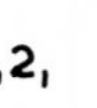

SIZES 17.
SMALL, MEDIUM, LARGE
LEAF TUBE

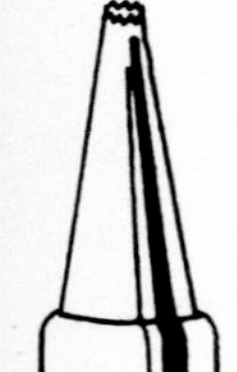

1. Pure icing sugar.
2. Jaconette or greaseproof paper bags.
3. Wooden spoon.
4. Turntable.
5. Various sized tubes (and screws if using Jaconette bags).
6. Glass or crockery basin (small).
7. Egg white and acetic acid.
8. Pair of scissors.
9. Knife and durex tape.
10. Set of decorator's nails and board, or waxed paper, for flower making.
11. Glass measuring jug.
12. Cake board and paper of choice for covering, usually silver or gold.
13. Strong, reliable food colouring of various shades. We have used Durchers and Raphael colourings throughout this book.

Icing Tubes and Pipes

No. 5, 8, and 13, small, medium, and large shell or star tubes.
No. 00, 0, 1, 2, 3, and 4 writing tubes.
No. 20 small, medium, and large petal tubes.
No. 22, basket tube.
No. 17, small, medium, and large leaf tube.

These tubes are sufficient for the beginner, extras may be added after practising and becoming competent with these few.

We have used Mathews' and Durcher's tubes, these are obtainable from all leading stores and most health food shops.

Care of Tubes

Always wash the tubes immediately after use; a small brush, millet straw, or feather, may be used with success, to clean the icing from the points of the tubes.

Rinse thoroughly under a strong flow of hot or cold water, dry well and put away; it is never wise to use sharp metal objects when trying to remove old dried icing from tubes as

* *"S", "M" or "L" (first letter) means small, medium or large. "R" or "L" (second letter) means right or left-handed worker.*

the risk of damage by this method is great. It is wiser to soak the tubes in water overnight.

Greaseproof Bags

The paper bag or cone has been found to be a clean, light and easy means of decorating, as well as being economical, as the bag may be cut in various ways thus eliminating the need for extra tubes such as the leaf tube and various sized writing tubes.

The size of paper for making the small bag is a triangular piece $6\frac{1}{2}$ x $8\frac{1}{2}$ x $10\frac{1}{2}$ ins. approximately.

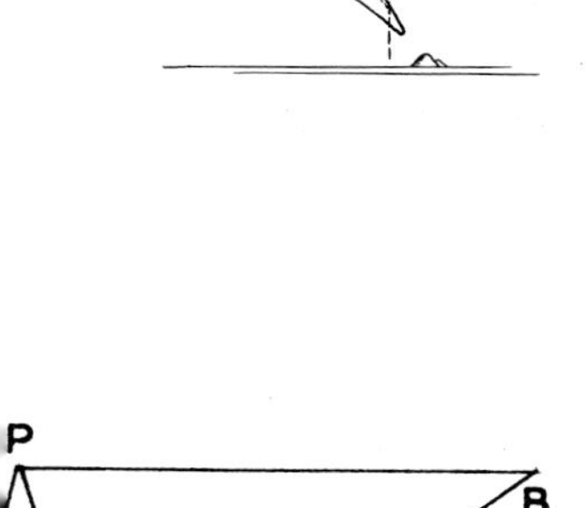

To Make Bags

Holding point A between thumb and index finger of left hand, grip point B with the same two fingers of the right hand and wrap twice around the left hand, over the thumb and first two fingers, until a sharp point is obtained at "P" (see sketch) and the lines PA and PB are concurrent.

Tuck the extending point of B back into the bag making sure there is a spear like point at the apex of the cone.

The bag is now ready to fill.

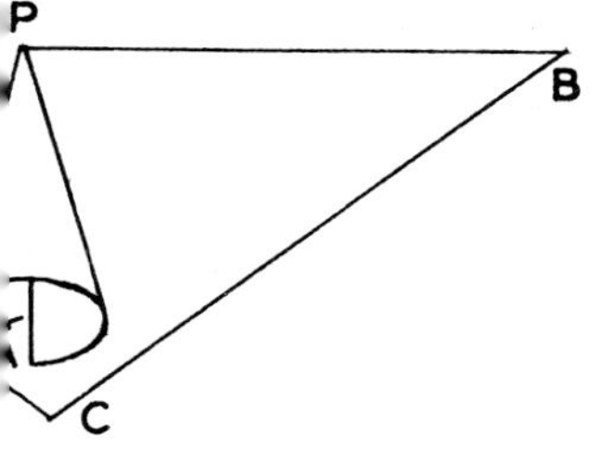

Hold the bag in left hand making sure the folded flap remains to the top of bag. Take small amounts of royal on the point of the knife and insert into the cone, being careful not to fill the bag too full.

Fold the top of bag neatly over and cut a small hole at the point.

The bag is now ready for use.

The large bag is made from paper measuring 22 x $16\frac{1}{2}$ x 13 ins. approximately, in the same triangular shape as the small bag.

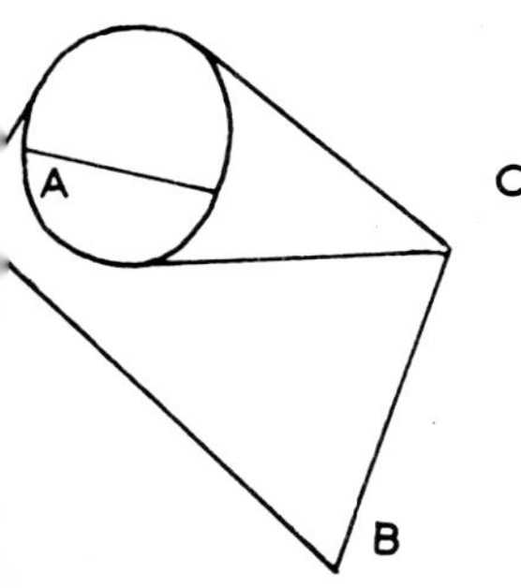

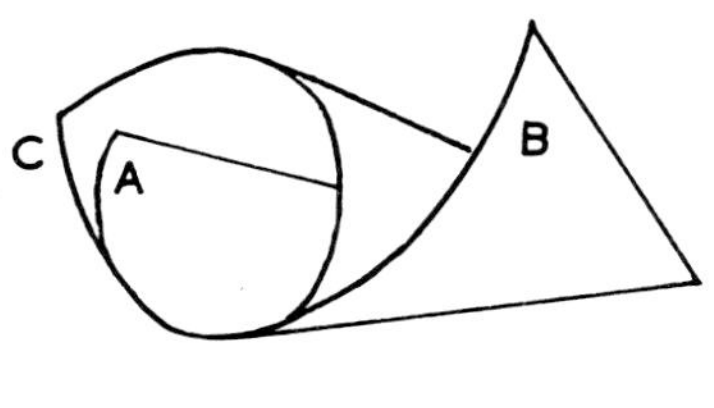

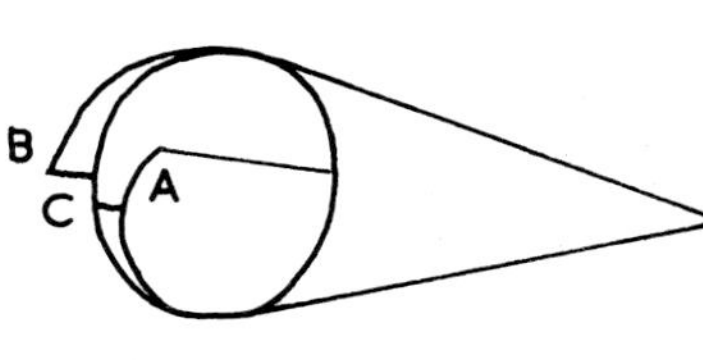

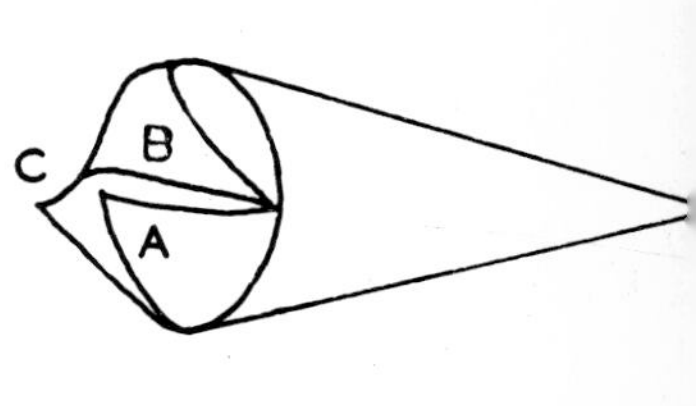

When the large bag is made, cut about $\frac{3}{4}$-in. from the point of the bag and insert the tube so that there is approximately half the tube protruding from the bag.

To fill the bag, hold it firmly in the left hand, dropping the royal into the bag from a wooden spoon held in the right hand.

Fold the bag neatly at the top, and it is ready for use.

Always keep the tops of tubes and points of small bags covered with a damp cloth when not in use as this prevents the icing from hardening.

The Making of Jaconette Bags

A quarter of a yard cuts into four 9 in. squares. Fold each square diagonally across the centre so that the rubberised surface will remain outside (see fig. 1). Join two of the matching sides with a double row of machine stitching leaving the remaining two open (see fig. 2).

Take the screw holding the thread end between thumb and index finger. Insert down to the point of the bag and tie securely with strong thread around the upper deep groove (see fig. 2 inset). Cut off the surplus point to the level of the screw. Turn the bag inside out, and it is ready for use.

Turntable

The turntable is a very necessary part of every icer's equipment. There is no need to go to great expense, as the home made types are just as effective as any other.

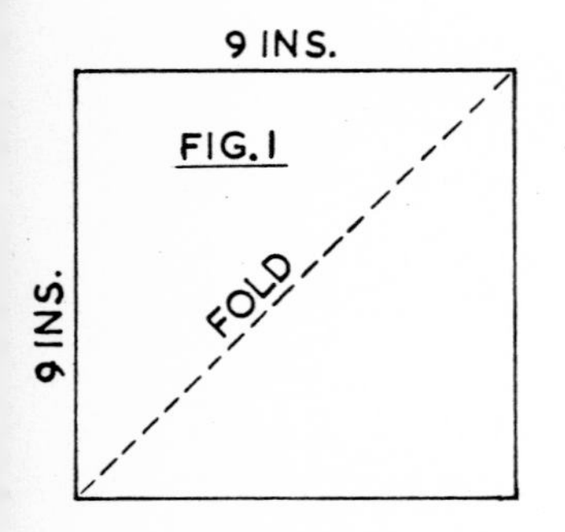

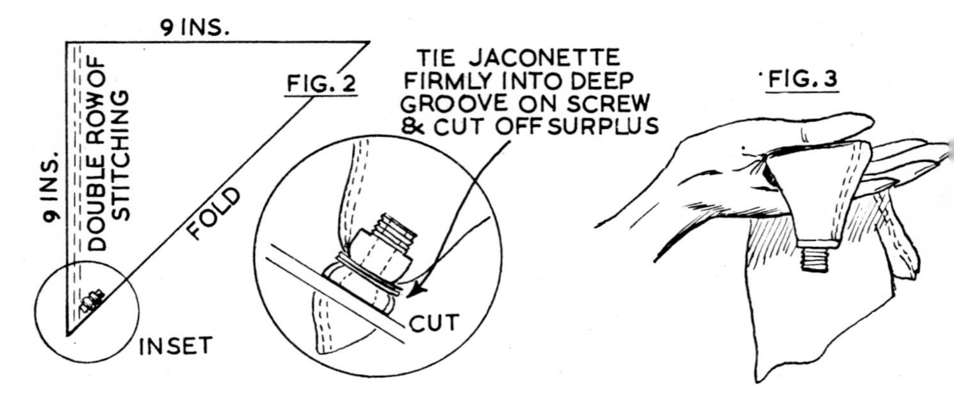

Royal icing and its uses

Royal Icing

White of one egg
6 to 9 ozs. finely sieved pure icing sugar (amount varies according to the size of the egg white)
2 drops acetic acid or a few drops lemon juice

Method

Sieve icing sugar through 32 gauge coffee strainer; place the egg white in a medium sized basin and slightly break through with wooden spoon. Add the icing sugar gradually to this, beating well after each addition until the consistency is such that it will form a peak on the back of the spoon upon being pulled from the mixture. Add 2 drops of acetic acid and beat; this helps to make the royal more pliable.

The success of royal icing depends upon the beating. The hand beaten method gives better results than an electrically beaten mixture as this tends to aerate the icing too much.

The royal may now be tinted to the required shade by adding any reliable brand of food colouring carefully with an eyedropper or skewer.

Royal icing is used for all pipe work. Never allow this mixture to be exposed to the air. Always store it in an airtight container and during use, keep it covered.

Food colourings may be bought in liquid or powdered form, the liquid being probably the best known and most widely used by the home decorator.

Always use colour in moderation.

The essential colours are: rose pink, burgundy, scarlet, green, yellow, orange, brown, sky blue, purple. Brown may be obtained by the use of parisian essence (a gravy colouring).

Many other colours may be obtained by blending shades together such as rose pink with a touch of yellow and orange will give a salmon pink. Lavender colours are produced by blending burgundy and blue. A drop or two of scarlet added to brown will give an improved colour. Greens are always improved by a touch of yellow or blue.

Be sure when purchasing colours that a strong brand is obtained otherwise if the cake is to be kept for any length of time the colours may fade noticeably.

Always check the consistency of the royal after the addition of liquid colouring as it tends to soften the mixture a little, and more icing sugar is then required.

THIS NOVELTY CAKE *is baked in an 8 in. tin. The small table and vase on top is all moulded work and rest on a piped variegated rug, accompanied by three moulded cats. The surface of the cake has been brushed with diluted brown colouring to represent a stained wooden floor.*

This attractive novelty illustrates the skill of arranging beautiful moulded roses and leaves artistically in a flower pot.
This excellent piece of work by courtesy of Mrs. D. Scott.

This Australian motif is ideal as a Bon Voyage, Welcome Home, or a Christmas cake for overseas friends or relations.
The motif is done in flood or cut-out work, and the detailing is hand painted when the work is dry. Fine extension and lace with ribbon insertion make an attractive border design, the 00 tube being used for this, and the No. 3 to build out the scallops. An assortment of Australian wild flowers form a pretty spray; these include the moulded Christmas Bells, flannel flowers, Christmas bush with piped wattle, and fern.

Above — *This simple little cake is applicable to many special occasions; it is so easy to decorate that the beginner should not hesitate to attempt the assembly of the moulded apple blossoms. These contain five finely fingered petals which are positioned in a firm royal icing to which flower stamens have been added. Small pieces of ribbon with piped sprigs of embroidery finish sides.*

Right — *A Golden Anniversary Cake, in pale cream covering and all over designed embroidery is sure to delight the happy couple. Golden coloured moulded roses, leaves and lily of the valley sprays set this cake off to full advantage. A small four petalled flower is free-handed over surface.*

This adorable puppy is another first prize winner. Done in flood work and touched up with the paint brush to bring out the vital details, this little fellow would delight the heart of any child on his birthday.

Sprays of piped flowers soften the front of the cake, and the candles are arranged around the board.

Animals such as this may be found in any child's picture book.

Another Show winner, this birthday cake has a soft, piped, embroidered pattern all over the sides, finished with a fine lace edge and extensions. The scroll is surrounded by a spray of assorted moulded flowers to make a pretty top. The flowers include fuchsia, lily-of-the-valley, roses and bouvardia.
The writing and key are painted over in silver paint.

An attractive 21st Birthday cake made 9 ins. square. The embroidered base was freehanded with the 0 tube. The ribbon insertion and base were piped with No. 8 shell tube. The embroidered pattern is carried on to the top, and the key is flooded and painted gold. Two blue birds hold writing. Sprays of lily-of-the-valley, hyacinths, bouvardia, and piped fern, surround rose.

Softly embroidered in freehand style with the 0 or 00 tube, and finished with No. 8 tube in stars for base, this 21st or Special Occasion cake is 9 ins. square.
An open book for the writing and key is surrounded by an assortment of moulded flowers.

This classic Special Occasion cake is 9 ins in diameter.
A pretty flooded floral design surrounds a large open rose and sprays of wired lily-of-the-valley. Moulded rose leaves, ribbon and tulle make a daintily finished top.
The built-out scallops at the base of the extension were piped with the No. 3 tube, and lattice lines and lace with No. 00 tube.

A nursery rhyme come true. This cute Humpty Dumpty atop his wall surveying the king's horses and king's men, is truly a little boy's party delight.
The grass effect in this novelty is obtained by dyeing coconut with food colour. The king's men and horses were store bought.

This novelty Wishing Well, is just the thing for any youngster's birthday party. The ducks, squirrel and fairies' chair are moulded. These may be replaced by the decorator who finds the modelling too difficult by the use of bought ornaments such as the frogs. The roof is cut in cardboard, covered underneath with silver or gold paper, and piped on top with variegated colours with the No. 8 shell tube.
Wisteria forms the trailing vine across the roof, which is supported by wooden skewers pushed into the cake. The lines of the brickwork may be painted on or marked with a knife.

A close-up photograph of birthday cake scroll and spray shown on page 21. A great deal of careful work is involved but the result is more than satisfying.

This lovely "Sweetheart" cake is just perfect for the Engagement Party. A large moulded rose on one stem accompanied by a bud and leaves, together with clusters of Lily-of-the-valley resting upon a dainty tulle handkerchief, make a delightful combination. A piped four petalled flower and embroidery fall softly over the sides; beneath these, the ½ in. space separating the two narrow bands of ribbon is filled with small royal icing dots. The base is finished with a No. 8 tube to form stars which are then interlaced with scallops.

For the 21st with a difference, we have chosen the one below. The moulded orchid spray with bouvardia and maiden hair fern, looks real enough to be worn as a corsage.
The scalloped extension, embroidery and lace pieces can be seen in clear detail, to give this cake pride of place on any birthday-party table.

A delightful design to open the door to much happiness. This is a simple cake, with its flooded door and wall, decorated by sprays of piped wisteria. The key is piped and painted gold, as is the name. The base has an extension, in a simple scallop banded by a narrow ribbon, to complete side design.

For this 8 in. 21st birthday, the fan is flooded on to waxed paper, allowed to dry and placed into position on the cake. A cluster of moulded roses, lillies, fuchsia, and tulle leaves, accompany a tulle piped handkerchief.
A detailed close-up of the side design can be seen below. A No. 00 tube was used for the embroidery and the No. 8 for the star base.

Our floral arrangement at base of fan on Mothers' Day cake illustrated in colour on page 32.

Mothers' Day cake with its beautiful piped fan and clustered spray of moulded violets, fuchsia and bouvardia, softened by embroidery and lace pieces, would delight any mother's heart. The base has been kept simple, just a shell separated by a star with a single scallop just above. The use of cut ribbon and embroidery is an added feature to the side design.
The arrangement at the base of the fan is shown on page 31.

*A pyrex dish was used to bake this pretty novelty cake. The top was then cut in an uneven shape and covered.
The moulded sweetpeas and leaves, make a good display.*

Pipework, Flowers and Sprays

ROYAL ICING is used for the piping of various types of flowers:— roses, sweet pea, daffodil, daisy, and forget-me-nots, etc. For these, the royal is used at a firm peak consistency, that is, a peak that will pull to a sharp point and remain erect. Flowers made from royal icing lend themselves to all types of decorating, from the novelty cake to the sophisticated wedding cake. Roses are perhaps the most popular and widely used flower, and if each of the following steps is repeated carefully, making them will prove no obstacle to the beginner. Using the large or medium 20 petal tube the rose is gradually built up around the point of the nail or toothpick.

To commence this flower, slightly grease the tip of each nail with Copha or margarine, then take it in the left hand at the end of the first joint of the index finger and grip the left side of nail with the right side of the thumb. Now roll the nail so that it travels along the full length of the first joint. Practise a few times, then place the nail back to the starting position, and, holding the bag in the right hand, place the tube along the right hand side of the nail so that the tube remains slightly higher than the nail point. The wide part of the tube should be at the lower end and the narrow part at the top, the concave side of the tube should face you. Commence to roll nail with the left hand, at the same time squeezing royal from the bag held in the other hand. This should enable the royal to spiral

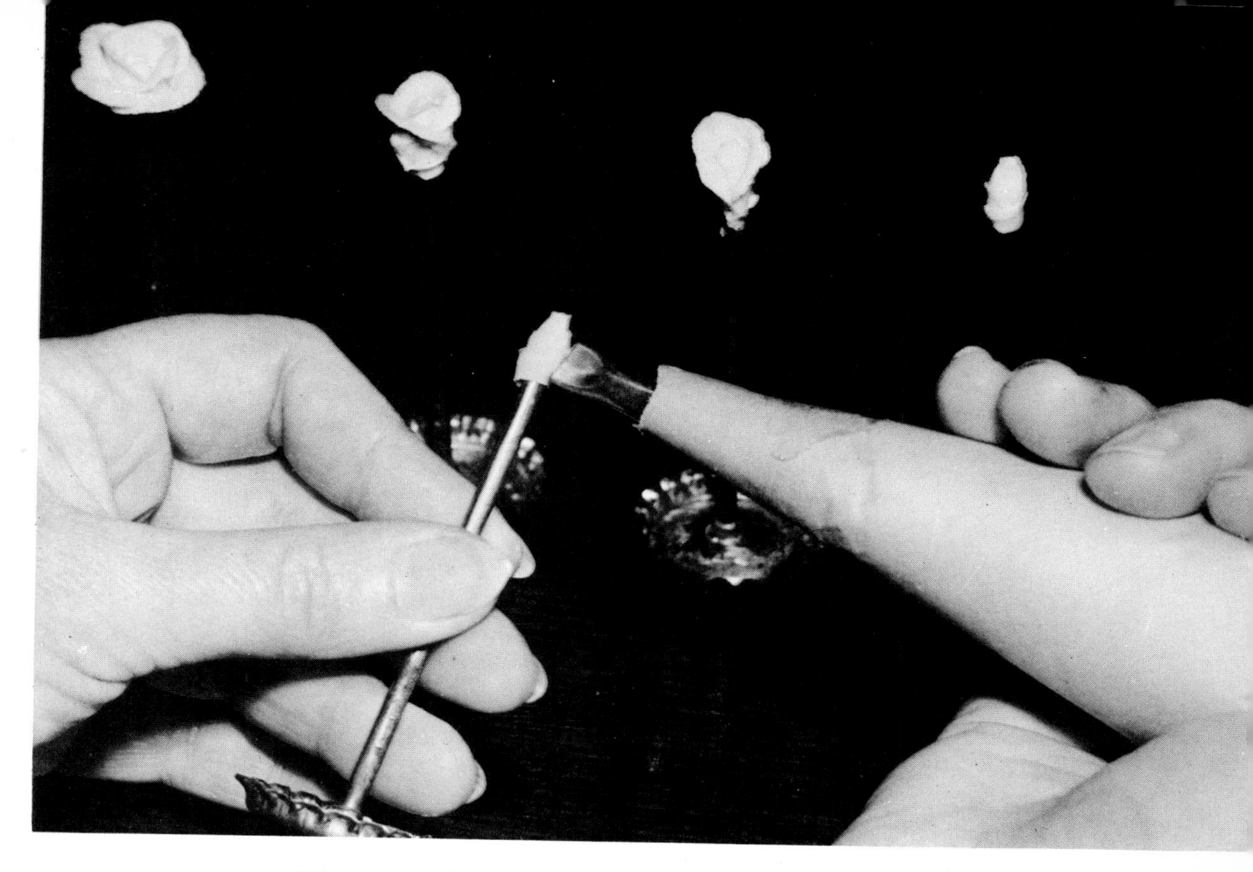

These are the steps in piping a rose, from the commencement of the bud to the finished flower. Note the position for holding the nail and bag.

around the top of the nail for approximately $\frac{3}{8}$ ins. This forms the bud base around which is built the rest of the rose.

Pipe all the buds first. For instance, if 12 roses are required, pipe the 12 buds on the nails throughout before you start to build the petals. Commence on the first and continue through to the last for each section. This allows the work proper time to firm before the rose is completed.

Keeping the lower edge of the tube close to the work, pipe an up and over movement in the shape of a semi-circle, thus forming the first petal, pipe another exactly the same as the first, and after this, a third one, turning the nail to correspond with the amount of royal being squeezed from the bag until all three petals have been piped. These three petals should remain slightly higher than the bud.

The next row is piped in the same manner, except that there are four petals, or if a larger rose is required, five. When the icing has firmed remove the rose from the nail and place on waxed paper to finish drying. The centres of these roses look most attractive if piped in a deeper shade than the rest of the petals. The piped leaf for these may be made by using the 17 leaf tube, or by cutting a small bag into a mitred point, and holding it in an upright position with the point of the bag just

This photograph shows the correct positioning of nails and bag for piping flowers.

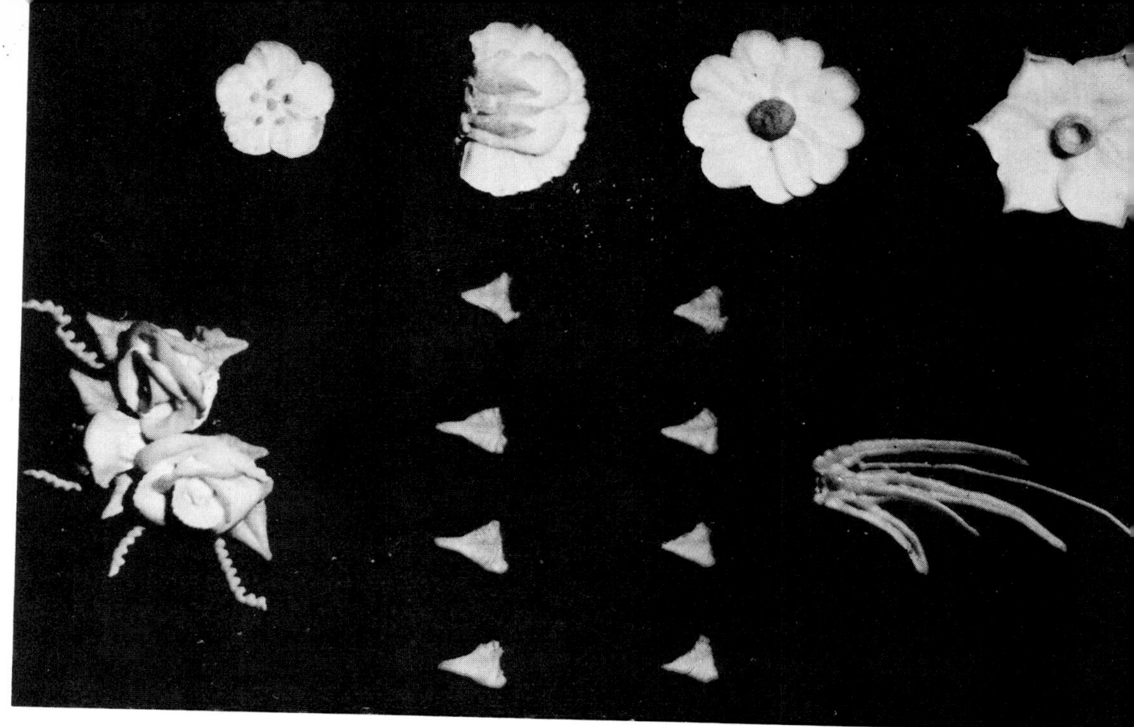

Some of the wide variety of flowers and small and the long leaves which the decorator learns to build up.

touching the surface of the cake. Squeeze firmly and then release pressure, tapering the royal off to a sharp point. Very often the paper bag gives a better shaped leaf than the tube.

Apple Blossom

This versatile little flower is one that is not too feminine to use on a man's cake. Pipe the flower directly on to a bottle top which has been slightly greased, or on to waxed paper. Pipe five small, semi-circular petals. Stamens may be added to the centres for effect or alternatively, you may pipe five small dots clustered around a centre one, using the small bag.

The Daffodil

Piped directly on to a greased bottle top or waxed paper. This flower has six petals piped with the medium or large 20 petal tube. Commencing from the centre of the bottle top, pipe a petal with a movement outward from and back to the centre

Grape designs are particularly useful as side decorations. Instructions for making them are given on the opposite page.

making sure that, by the time three petals have been piped, there still remains sufficient space to complete the last three petals. Using an 0 or 1 writing tube, or small bag, pipe the trumpet by building approximately five rows on top of the other until the centre is the required height. Before the six petals are dry, pinch the tip of each between the thumb and index finger, thus changing the rounded look to a pointed one.

Daisy

Pipe in a manner similar to that used for the daffodil but on a smaller scale, as there are approximately 10 or 12 petals to be fitted on to the bottle top for the daisy. When piping the last petal be sure to raise the tube slightly so as not to damage the first petal. A centre dot may be piped or stamens added.

Sweet Pea

Using the same tube as for previous flowers pipe, with a steady pressure, a half circle on the top edge of the bottle top, at the same time turning the nail to correspond. The second petal is piped in the same manner but with two halves inside the top circle. Holding the bag in an upright position in the centre of the flower, pipe an out and back movement, and repeat this on either side to form three centre petals. Allow to dry on the bottle-top and remove flower by holding over dry heat.

Grapes

Using the small paper bag or the No. 0 or 1 writing tube, these may be piped on to waxed paper or directly on to the cake. The consistency of the royal should be a little firmer than is required for lattice work.

To pipe the grape squeeze a base the size and shape of an almond. Holding the bag in an upright position, commence at the narrow end of the base and force out a shape similar to a teardrop, starting with one, for the first row, two for the second row, three for the third, and so on until the entire base is covered with teardrops. If the last row is reduced back to three teardrops this will give a rounded look to the fruit. And reduce the possibility of making the top too broad. Slightly overlap each dob of royal keeping in mind the triangular appearance of a bunch of grapes; the leaves and tendrils are piped to the grape when it is placed into position on the cake. These decorations are ideal for ornamenting masculine cakes.

Wisteria

Pipe with a base the same size and almond shape as described above for the grapes. The top petals may be piped with a small paper bag, which has been cut in one side only and at an angle, holding the bag in an upright position with the cut edge pointing away from you (see sketch page 11). Pipe a short line forward and then return back over the forward line. This forms a small fluted petal and is piped individually from one at the base increasing in number until the almond shaped base mound is completely covered and looks a good shape. For those wishing to use a pipe, the 20 small petal is recommended and held in an upright position; to maintain a good shape and not broaden the flower too much it is wise to reduce the last row back to three petals as for grapes.

These flowers are piped on to waxed paper and allowed to dry and are then placed into position on the cake with a little royal icing. Leaves and tendrils are then added.

Wired Sprays and Fern

All the under-mentioned wire sprays are cut and piped in a similar manner, the fine wire is cut in short lengths of three or four stems and twisted well at the base to prevent them separating. These are then placed on to a board which has been covered with waxed paper and the pipe-work is done directly on to the wire stems and allowed to dry before removing.

WATTLE. Colour the royal icing a golden shade and using the No. 1 tube commence to pipe a few small dots at the top of each wire, then form clusters of three dots along the rest of the wire. Increase the pressure towards the end to give larger blooms, allow to dry and then remove from the waxed paper.

FERN. Using a No. 0 tube, colour the royal a deep rich green and pipe a series of elongated dots along each side of the wire stem, slightly increasing the pressure towards the base.

MAIDEN HAIR FERN. Dye the wire stems brown or black with a little of the food colour, and using a very pale green icing commence to pipe a few small dots at the top of each wire and then pipe clusters of three dots, increasing size towards the base. Before the royal icing dries, take a separate piece of waxed paper and place this over the top of the fern and gently flatten out the clusters. Remove top piece of waxed paper from fern and allow to dry.

FORGET-ME-NOTS. With a small bag or No. 1 tube pipe a few dots at the top of each wire stem, then make five small petals directly on to the wire to form a very small flower and finish it off with a tiny dot of a contrast colour in the centre of each. Continue in this manner until stem is covered.

Basket Weave

Basket weave as a decoration is applied directly to the cake. It is used extensively in novelty work such as baskets of fruit or Christening cakes shaped like a bassinette.

Using the No. 22 basket tube or the shell tubes, 5 or 8, work a horizontal line around the top of the cake. Then make short vertical lines at regular intervals over the first line, extending the vertical line at least $\frac{1}{8}$-in. longer than the width of the horizontal one. Follow with another horizontal line close enough to cover the extension of the vertical one.

The second row of vertical lines is placed between the spaces of the first vertical lines. Succeeding lines follow this pattern.

A lighter looking basket weave may be obtained by using any of the writing tubes from 0 to 4 for decorating small baskets of flowers for wedding or birthday cakes.

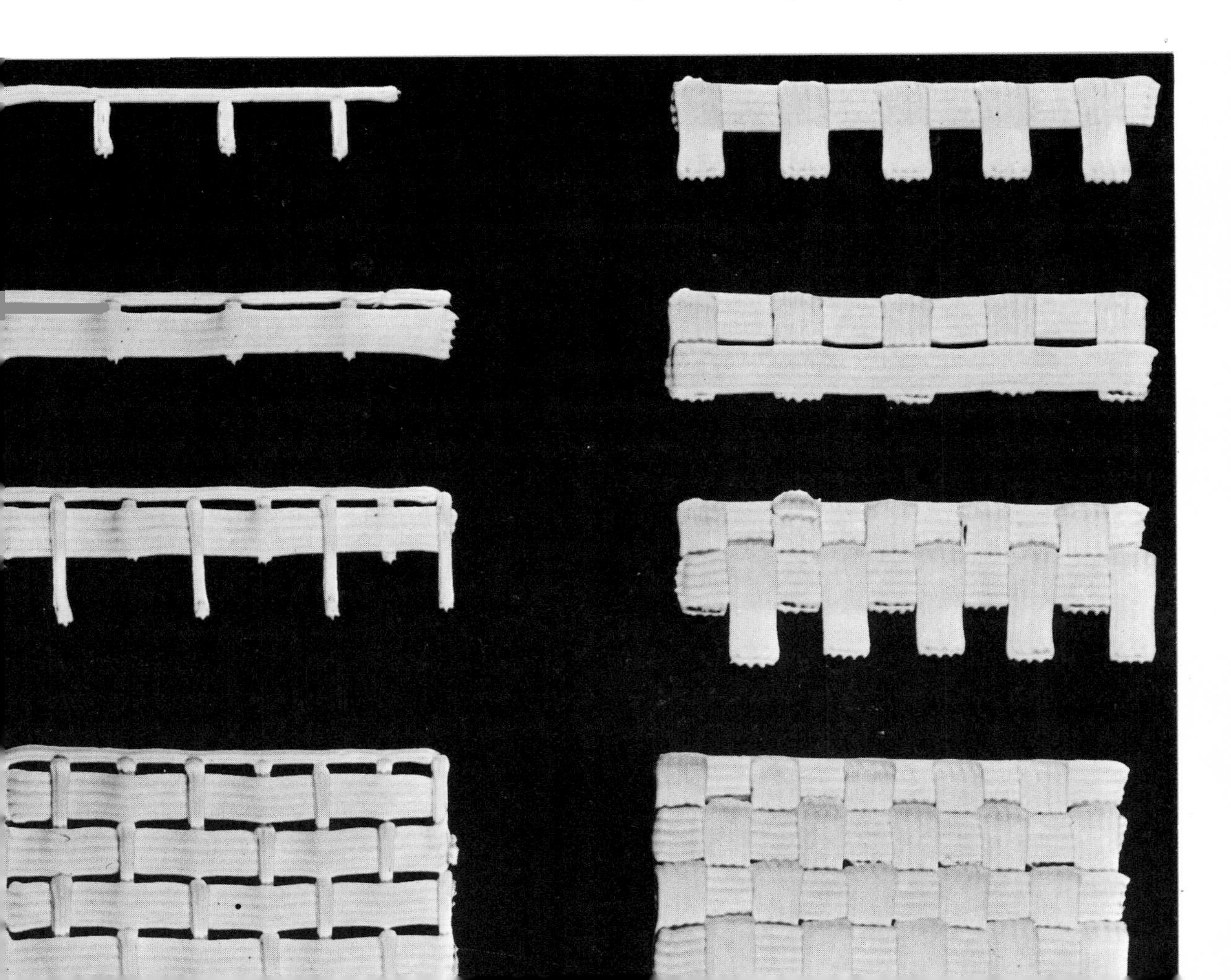

Easter Eggs

CHOCOLATE EGGS are usually the most popular, and provided they can be stored under, cool, dry conditions, there is no reason why the decorating may not be done well in advance of Easter.

Moulds for making eggs may be purchased in plain or crocodile pattern, but as these are comparatively expensive most people choose to purchase plain chocolate eggs from the confectioner.

This dainty easter egg with its beautiful moulded flower spray, would be a delight to young and old alike.

Easter egg time is mainly for the children, and this little bear has loads of appeal. The design may be done in flood work and coloured as desired, or in cut-out work. A spray of moulded flowers helps to complete the design. Any motif from a book or card may be used in the same way.

Cake covering and preparation

THERE ARE VARIOUS fondants and plastic icings for cake covering, that can be used by beginners as well as more advanced decorators, with assured success. The smooth satin-like finish is a delight to behold and keeps for an indefinite time, yet remains pleasantly soft to eat.

The two recipes below are strongly recommended for success.

Plastic Icing

To cover approximately a ½-lb. cake (7 in. or 8 in. cake tin)

2 lbs. pure icing sugar (sieved)
4 ozs. liquid glucose
½ oz. gelatine (level tablespoon) (such as Davis Finely Ground Gelatine)
¾ oz. glycerine
2 ozs. water
Flavouring to taste, personal choice.

Method

Put gelatine and water in saucepan, place over gentle heat, stirring constantly, do not boil. When gelatine is completely dissolved, remove from heat. Add glycerine and glucose and stir. Add this to sieved icing sugar in basin and mix with hands to firm dough. Add flavouring. Place on a board lightly sprinkled with cornflour and knead very well.

Carefully add desired colour with eyedropper or skewer and knead again until colour has been evenly distributed through the mixture. Roll out to uniform thickness and size required to cover the cake. If not intending to use immediately put into an airtight container or plastic bag. Do not place in a refrigerator as this tends to harden the glucose, requiring much kneading to get it back to a pliable consistency again.

Plastic Fondant

1 lb. crystal sugar
¼ lb. liquid glucose
5 ozs. water
1 oz. glycerine
1 level teaspoon cream of tartar

1 oz. gelatine
5 ozs. hot water (to dissolve the gelatine)
¼ lb. Copha
3 lbs. sieved icing sugar

Method

Grease saucepan around top with Copha and boil the first five ingredients listed above to the 240°. When bubbles disappear add dissolved gelatine, never stir while boiling. Allow to stand about five minutes then add flaked Copha. Turn into bowl and gradually add icing sugar. Cover bowl with plastic cover and keep airtight. Allow to stand for 24 hours before using. Add extra icing sugar to obtain desired consistency.

Almond Paste

Almond paste base can be used to help give a smoother surface to the finished cake.

This recipe will make enough to cover an 8 in cake.

1 lb. pure icing sugar (sieved)
4 ozs. pure ground almonds or marzipan meal
2 egg yolks
2 tablespoons sweet sherry
Squeeze of lemon juice (optional).

Method

Mix almond or marzipan meal through icing sugar. Make a

well in centre of mixture in basin, add beaten egg yolks and sherry combined. Knead into firm dough. If too moist add extra icing sugar, if too dry add extra sherry.

Brush cake with egg white or the puree of a seedless jam. Roll out almond paste and place over cake. Trim off surplus from base of cake and rub hands dusted with icing sugar over surface of cake. If using marzipan paste, allow to stand at least 12 hours before covering with plastic icing, whereas an almond base would require approximately two days drying in order to prevent the almond oil staining the cake cover.

Preparation for Covering Cake

The cake should be a fruit mixture or a good butter cake, but not of a light nature such as a sponge. Should the cake rise to a dome it is advisable to level that piece off and so save a lot of packing, assuming that it is intended to decorate the bottom up of the cake, which gives a perfectly flat surface on which to work.

Some designs call for the natural rise to the cake's surface. If your design is such, leave the slight dome and lightly bevel the top edges to give a rounder appearance.

Take portion of the plastic icing or almond paste and commence to pack any fruit holes or paper creases, smooth each piece of packing off with the back of a knife, so that the packing does not remain higher than the surface of the cake. Work carefully for a smooth finish.

If it is a square cake, make sure the four corners are exactly square and the same height. If uneven, they must be falsely packed. After the cake is completely levelled, it is then brushed all over with the white of egg or the puree of a seedless jam (apricot is one of the best). If the jam seems too thick for brushing, thin it down with a little hot water. Make sure the jam is only lightly applied. All this should be done before the plastic icing is rolled out.

If the cake has been previously covered with almond paste

the surface should be brushed over with egg white or jam before covering with the plastic icing.

Knead the plastic icing well before rolling and dust the surface lightly with icing sugar or cornflour, so that premature sticking will be avoided.

Roll out to approximately a $\frac{1}{2}$-in. in thickness and place over the cake with the use of a rolling pin.

Dust the hands lightly with icing sugar or cornflour and rub first over the top surface of the iced cake, then over the sides until a smooth, even surface is obtained. Don't keep rubbing unnecessarily. Cut the uneven edges away at base of the cake with a sharp knife and place on a covered board.

The board should be Masonite or a strong ply wood and should be $2\frac{1}{2}$ ins. to $3\frac{1}{2}$ ins. larger than the cake. Cover the board with silver or gold paper or paper of personal choice. The cake is now ready to decorate. Try to place the design with such exactness as to ensure that the finished cake will look properly balanced as well as attractive.

Francine

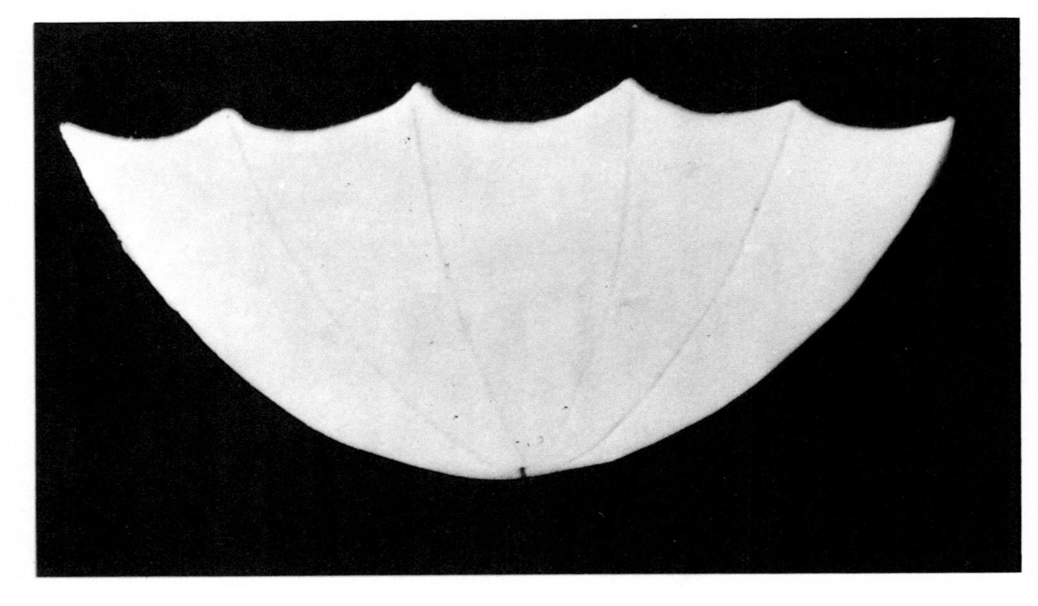

Above — *Orchids and lily-of-the-valley grace the top of this birthday cake which has a small all-over flooded pattern, and soft embroidery to finish off the side design. Piping is done with No. 0 or 00 tube and No. 8 for star base.*

Above Left — *Another simple cake for birthday or special occasions, the moulded umbrella is shaped over waxed paper until dry and daintily filled with roses, bouvardia, for-get-me-nots, and green fern. The handle and name is piped and painted gold. The butterfly is piped on tulle and assembled into a piped body of royal.*

Below Left — *This umbrella for the birthday cake above is cut from modelling paste and shaped over a dome of waxed paper to obtain the curved look. We have used this on our girl's birthday cake "Francine", and filled it with flowers.*

This lovely birthday cake is adorned with large, open, full-blown rose, lily-of-the-valley and bouvardia, with a little tulle to soften the spray. A small flooded leaf pattern adorns the top and sides. Decorated in golden tonings it could also be used as a Golden Wedding cake.

A Victorian posy of roses and forget-me-nots, make this dainty design suitable for use as a debutante, anniversary or birthday cake. The lace and extension work have been done with the 0 or 00 tube and the floral pattern was piped freehand.

Net Work

NET OR TULLE gives an added touch of softness to the finish of a cake, and may be used in the form of such decorations as bows, handkerchiefs, veils, leaves, frills and straight lace designs. Royal icing used for piping on to net should be at a soft peak consistency. That is a little softer than that used for flower work. A rayon or cotton net will require the use of a stiffener, and the following recipe is ideal for the purpose—½ cup of pure icing sugar, ½ cup of water.

Method

Place sugar and water in a saucepan and stir over gentle heat until sugar is dissolved. Bring to the boil and allow to simmer. This will keep indefinitely if stored in an airtight container. To stiffen the net immerse pattern pieces in the solution and shake the net well to free the mixture from the holes. To dry place on a flat surface. A nylon tulle gives a more delicate effect than cotton or rayon net and does not require to be stiffened.

Net Bow

Using the 00 or 0 tube, place the cut bow pieces on waxed paper on a flat surface. Pipe a teardrop, or snail-trail finely around the outside edge. Inside this do a line of small scallops about ¼-in. from the teardrop edge. Fill in the remaining area of net with scattered dots, or a design of personal choice. Place the two bow pieces over a broom handle or piece of dowell to dry. Tail pieces must be made as a pair and are piped on waxed paper in the same manner and design as the other two pieces of the bow. The small centre piece is piped with a tear-

Steps in making a tulle bow — the pieces of tulle, half the bow made up and the completed bow. Also tulle butterfly, set out in sections and assembled. The butterfly's wings may be varied in shape and design, and piped over tulle on waxed paper. When dry remove the paper and pipe the body with No. 8 shell tube on a small piece of waxed paper. The wings are then placed into the wet royal body and supported by some small object until they have dried into shape. The feelers may be piped or flower stamens used.

Detail of piping an ivy leaf on waxed paper and arranging the leaf to look like a trailing vine. Also, the piping of rose leaves on tulle. The embroidered handkerchief may be used on the corners of wedding or special occasion cakes, or on the top with a spray of flowers to provide a softening touch.

drop edge and a series of dots over the net. Allow this piece to dry over the handle of a wooden spoon.

To Make Up

Remove the pattern pieces from the waxed paper. Place a good dob of royal approximately a ¼-in. in diameter on a separate piece of waxed paper and secure the under edges of the bow into this. Squeeze another dob of royal about the same size on top of the previous one and use this to secure the two tail pieces into position. Squeeze another small dob of icing on top of other two and fold the top edges down into this. The small centre piece is then placed above this area. Allow to dry thoroughly before removing from paper (see illustration page 53).

Handkerchief or Bride's Veil

Cut a 7-in. circle or square and fold about two thirds from the top. Fold the side pieces in towards the centre, giving the appearance of a fichu. Pipe around the edges of folds with a teardrop, snail-trail or small scallop as previously directed for the bow, and design with any fine embroidery such as shown in the illustration on page 53.

The Ivy Leaf

A favourite for softening the lines of a wedding cake, these leaves may be varied in size from the large leaf to the very small one, and arranged like a trailing vine.

They also are piped with 00 or 0 writing tube, or small paper cone using the small teardrop or a scratchy irregular stroke around the outside edge, and worked over waxed paper. If the leaves are to be curved, dry them out over curved objects such as a rolling pin, broom handle or dowel rod for smaller leaves.

Rose Leaf

If requiring the flat leaf, these may be piped in numbers directly on the net over waxed paper and cut out later with a small pair of scissors.

Using 00 or 0 writing tube or small paper cone, drop the outline of the leaf and pipe in the veins. If a curved leaf is required, dry over a curved object as previously described.

This pretty cake is suitable for a wedding or 21st birthday. The scalloped lace design filled in with extension work achieves a dainty finish to the base of the cake. Ribbon threaded through a soft piping of embroidery, and an embroidered centre piece all add to the general appearance of the design. The floral sprays consist of roses, hyacinths, bouvardia, and lily-of-the-valley, tulle leaves add finish to the sprays.

Bill

Above *A special occasion cake for a Shower Tea. Bride and stairway were done in flood. Flowers consist of small piped roses, lily-of-the-valley forget-me-nots, and sprigs of fern piped on wire. Lace and extension trim the sides in a heart pattern piped with 00 Tube.*

Left *This delightful little bear is flooded directly on to the cake. Small piped flowers and blades of grass line the roadway. Suitable patterns such as this one may be found in any child's picture book or on birthday greeting cards.*

The Dolly Varden cake, ever popular for litle girls' party cakes or debutantes. The cake may be cooked in a special tin or a pudding basin. This design makes use of a circular tulle skirt placed over soft embroidery and extension work. The flowers used in the spray are piped rosebuds and forget-me-nots.

An adorable bride is used here for a little girl's birthday cake. It may also be used as a debutante's cake, without the veil, or for a Shower Tea party. The skirt of the doll is cut in a circle and allowed to fall softly in flares over the cake. Crimper work and cornelli add a feature to skirt and bodice with a lace edge.

Moulded Flowers

THE MOULDED FLOWER gives a very realistic touch to cake decorating. It is advisable to copy from a real flower or illustration, to ensure using the correct number of petals.

Care must be taken in this type of work to finger the petals and leaves finely, otherwise they can look heavy and clumsy.

Two successful modelling paste recipes are as follows:

1 lb. pure icing sugar
½ oz. liquid glucose
¼ oz. gelatine
2 ozs. water.

Method:

Place gelatine and water in saucepan and stir over gentle heat until gelatine is completely dissolved (do not boil).

Remove from heat, add glucose and stir. Allow to cool slightly, add to sieved icing sugar in basin and knead well until all icing sugar has been absorbed into the liquid.

Place on a board lightly sprinkled with cornflour, add colour as desired and knead well. Place in airtight container otherwise the air will dry this mixture off quickly.

If mixture seems too dry, this may be remedied by using a little extra water. If too moist add extra icing sugar. The consistency should resemble that of plasticine.

½ lb. pure icing sugar
rounded teaspoon gelatine
rounded teaspoon Copha
3 dessertspoons of water.

Method:

Sieve the icing sugar into a basin, dissolve gelatine, water and Copha over a gentle heat. Add to icing sugar and mix with a wooden spoon. Cover with plastic cover, and use extra icing sugar to bring to the desired consistency.

Open Dog Rose

This rose is formed with five petals. Taking a small ball of modelling paste between the cushions of the thumb and index fingers, press out thinly into the shape of a rose petal. Shape the top of the petal by turning it over your finger, and shape some others by curving in both sides slightly to give a dented effect. The size of the balls of modelling paste used to finger these petals, may range from a pea to a marble, according to the size required for the finished rose. If a more cupped rose is needed press the cushion of the thumb into the centre of each petal. These are then placed into round based patty tins and allowed to dry. When dry the petals may be tinted with diluted food colouring and painted from the base of the petal to approximately half way for the front of the petal *(illustration page 61)* and the back is painted with a deeper tone all over. To assemble, use a firm royal icing and a No. 5 or 8 shell tube. Place a piece

This lovely spray of dog roses, hyacinths and autumn toned rose leaves, make a pretty top for any birthday cake. Piped thorns have been added for effect.

Shape of the frangipani petal and the assembled flower. To mould a very open flower, the petals may be shaped over a piece of broom handle. These are lightly painted with diluted yellow colouring and allowed to dry. Paint only about two-thirds of the petal and assemble with royal, or as illustrated allow each petal to overlap the previous one and twist at base and allow to dry over a paper cone.

of waxed paper in the bottom of the patty tin and make a heavy dot of royal in the centre, attach the petals in the icing. Each should overlap slightly. A small star is then piped in the centre of the rose and stamens, cut about $\frac{1}{2}$ in. long, are added with the help of tweezers—allow to dry at least 12 hours. The leaves may

Showing the water lily assembled in patty tin, with the leaf drying to shape (note the little wedge cut-out).

Board showing the three different sized petals for moulded water lily before being shaped in patty tin.

be cut freehand or with the aid of a pattern drawn from a real leaf. They are lightly veined and set to dry over the edge of a board, saucer or rolling pin to obtain a curved look. The stems are made by rolling a small piece of modelling paste on to a board with the fingers and hands until a stem about the thickness of a wooden skewer is obtained. The length of these stems will be determined by the size of the spray and also the size of the cake. To make a bud press a small ball into a petal shape and roll from end to end, pinch at the base and cut off any surplus.

Jonquil

Stages in making-up a jonquil, with piped stems and leaves.

These flowers contain six petals. For each take a small piece of modelling paste about the size of a pea and press it out thinly to a petal shape and pinch to a point with the fingers. Allow them to dry over the handle of a wooden spoon. The centre of the flower needs a piece of paste about the same size as that used for the petals; this is shaped into a small cup-like piece over the end of a paint brush. Assemble the flower on a piece of waxed paper with a squeeze of firm royal icing large enough to cover a one cent piece, arranging three of the six petals with the broad ends to the centre, to form a triangle and the pointed ends to the outside as shown in illustration page 65. The remaining three petals are placed on top of the first three so that their outside pointed ends cover the spaces between those of the lower three. The trumpet piece is placed in the centre with a little royal and a few stamens, about the same height as the trumpet, are added. Mould or pipe the stems and leaves.

Christmas Bell

Take a ½ in. oblong piece of pale orange coloured modelling paste and shape it over the end of a size six or eight paint brush. Slit down to form six equal petals about ¼ in. to ⅜ in. deep. Recut each petal to a point with a pair of sharp embroidery scissors or pinch them with the fingers to form a scallop effect. Paint the outside of the bell a bright scarlet to within ¼ in. of the petals which should be left pale orange both outside and inside. The scarlet should not finish abruptly in a line straight across the bell, but instead, should be in scallops matching the shape of the petal edge. Stamens may be added.

Christmas Bush

Take a small piece of modelling paste about the size of a pea

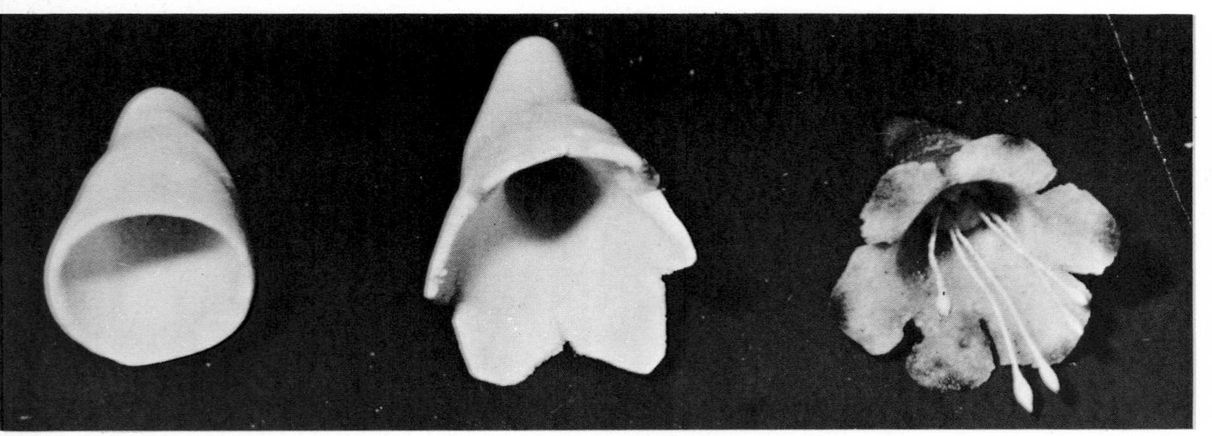

The three major steps in moulding a Christmas bell.

and, after shaping it over the end of a paint brush, as illustrated on page 68, slit it down to form five equal petals. Cut these into points with sharp scissors and insert a fine wire which has been hooked, or knotted, at the end. Pull the wire through the flower until the knot embeds itself in the base and finger it firmly. Add a bright yellow dot of royal icing to the centre of the flower.

Hyacinth

Use a small piece of modelling paste shaped over the tip of a brush handle as shown in illustration on page 68. Slit it to form six petals by halving and then by trisecting each half thus giving six evenly divided petals. Cut the petals to a point with a small pair of scissors and holding each between the fingers, turn it back slightly. Lightly press the handle of the paint brush into each petal. Wire flowers as explained previously. Centres may be painted lightly with diluted food colouring.

Assembly board photograph showing the stages of the hyacinth, boronia, and lily-of-the-valley.

Bouvardia

Shape a piece of modelling paste into a narrow roll about ½ in. in length. Insert the handle of a paint brush through centre to approximately ¼ in. from the end. Cut into 4 pointed petals and insert fine wire with a hooked top. Using a twisting, turning movement, work the paste firmly down over the wire until a tapered point is produced.

This flower and many others of its type may also be shaped by using the pointed end of a knitting needle.

Lily-of-the-Valley

A small piece of modelling paste is shaped over a brush handle and serrated around the top with tweezers or small scissors. Knot a piece of fine wire, and insert it down through the base of the flower a few minutes after it has been made and secure the wire with a little royal icing.

Rose

Form a cone or rolled piece for the centre. Take a small ball of paste and finger it out finely to a rose petal shape. Attach it with a little water to the centre piece. Two of these petals will probably complete the first row. For the second row, slightly increase the size of the ball of modelling paste and finger out finely, as for the first row, to the shape of a petal. Attach it to the rose and continue in this way fixing each petal to the centre of the previous one. Continue in this manner until the flower becomes a good shape. The full blown rose is moulded in the same way, except that the centre is pinched out and manufactured stamens inserted to give effect.

The Cecil Bruner rose in its various stages of assembly.

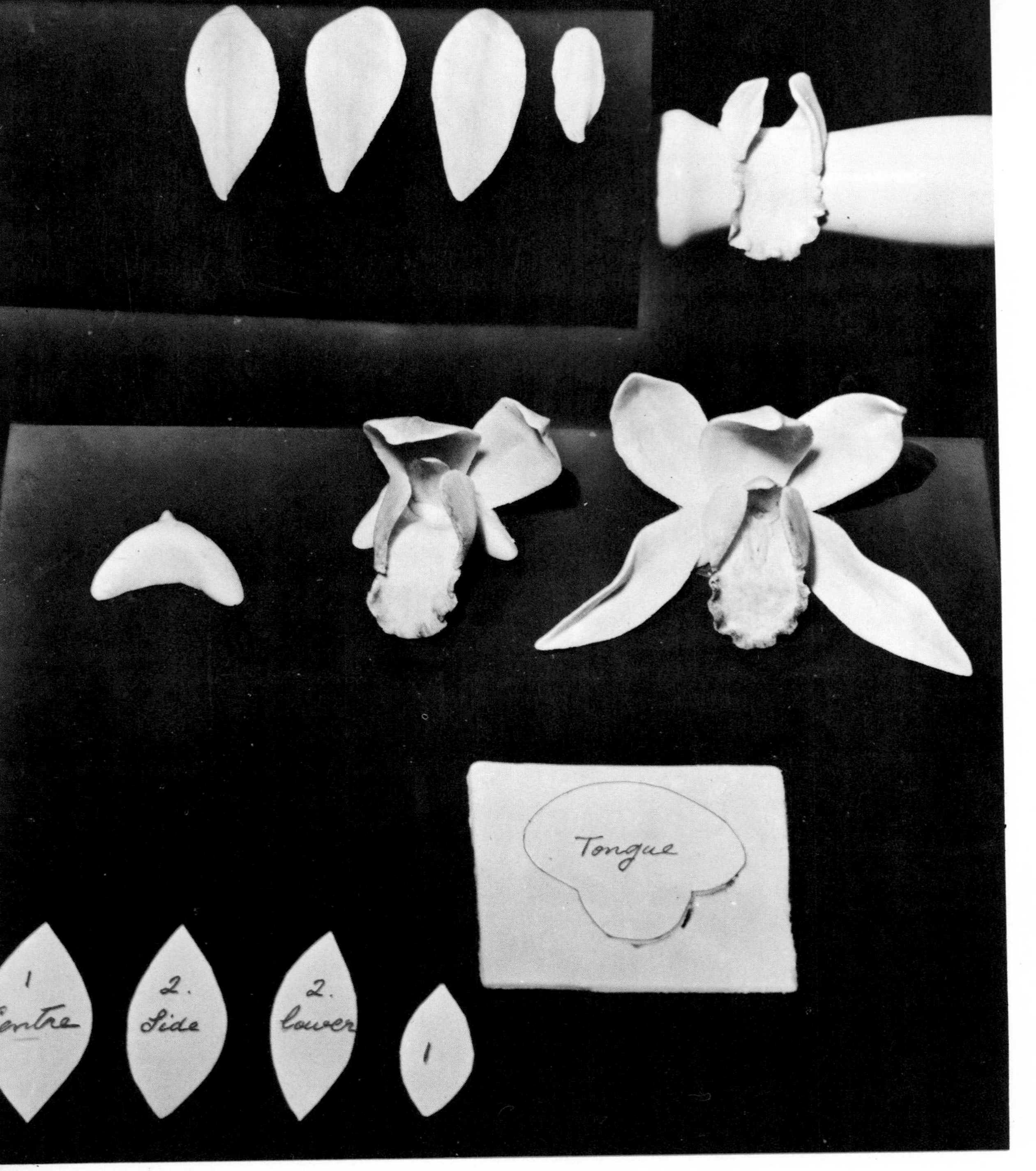

The orchid, showing the positioning of the pattern on the modelling paste for cutting out, and the assembly.

Orchid

Roll out a small piece of modelling paste and cut one piece of the pattern at a time so as to avoid having the pieces drying too rapidly. Cut each piece with a sharp knife and drape it over a rolling pin to dry, as shown in the photograph (page 70). The three top petals and small top tongue are dried over the roller part of the pin and the large tongue is shaped over the narrow part of the handle, with the side pieces turned up and the tongue edge slightly fluted. The two lower petals are twisted slightly and allowed to dry out flat on a board.

When dry, the pieces are coloured with a paint brush, and assembled into a banana shaped piece of paste by pressing each petal into position and attaching it with royal icing. Any surplus may be cut away from the base upon completion of the flower.

Boronia

The colours of this flower may vary. Deep burgundy outside petals would have a pale pink inside, whilst the brown boronia outside petals would be pale yellow inside.

It is wise to mould the flower in the pale colour of the inside and paint the deeper colour required for the outside when the flower is completed.

Mould a small piece of paste (the size of a pea) over the end of a brush as shown for the hyacinth. Cut four or five pointed petals and pinch them lightly between the fingers. Attach wire into the base of each, using a little royal icing. Add a few short stamens. The bud is a tiny ball of paste with a few knife marks on top to represent the unopened petals.

Waratah

Take a small piece of modelling paste and mould it into a dome shape. Pipe it all over with a series of dots, which become much larger as they near the base of the dome. Set this piece aside to dry.

Roll out some paste and cut eight or nine large elongated petals, and place them to dry in round-based patty pans. Give each petal a slight bend or twist to make the finished flower look more natural.

Repeat this process, cutting eight or nine smaller petals, and dry in the same way in patty pans. Repeat again, this time cutting nine very small petals to fit close to the dome section.

To assemble, place a piece of waxed paper in the base of a patty tin and group the three rows of petals around base of dome, the smallest being close to the dome, and the large petals forming the two outside rows.

Use a firm royal icing to put the flower together and allow it to dry in the patty pan in order to keep its shape. If you are unable to work modelling paste into a bright red or scarlet colour, the deeper colours may be obtained by painting the petals, and dome, when thoroughly dry, with scarlet food colouring.

The leaves are in a deep shade of green, veined with the back of a knife and the edges are pinched around with pointed tweezers to give a serrated look.

Sweet Pea

Take a small piece of modelling paste and finger it out to form a small petal. This is dampened with water and is folded in half around a short piece of wire. Take a slightly larger ball of

Finished waratah assembled in the patty tin, with the basic elements alongside, to indicate their relative size.

paste and press out a larger petal. Flute the top edge of this with the aid of a knitting needle. Slightly dampen the base and wrap it around the centre piece. The third petal is larger again and the top is fluted. Slightly dampen the base and wrap it around the outside of the previous petal and turn it back slightly.

To give more variation to the colours, lightly stroke them with food colouring from a paint brush. The leaf is long with softly fluted sides. These may be wired if required.

Violet

This dainty flower has five petals the centre one which is the broadest, is cupped into shape over the cushion of the thumb. The other four are shaped to make long petals and are lightly pinched down the centre of the back. These pieces are allowed to dry and are then painted a deep violet leaving a small white area in the centre.

The calyx is wired, and the petals are secured to this with a little royal. A centre dot of yellow icing finishes the flower. The leaf is almost round in shape and is veined similarly to that of the water lily.

Detailed photograph shows the sweet pea in assembly.

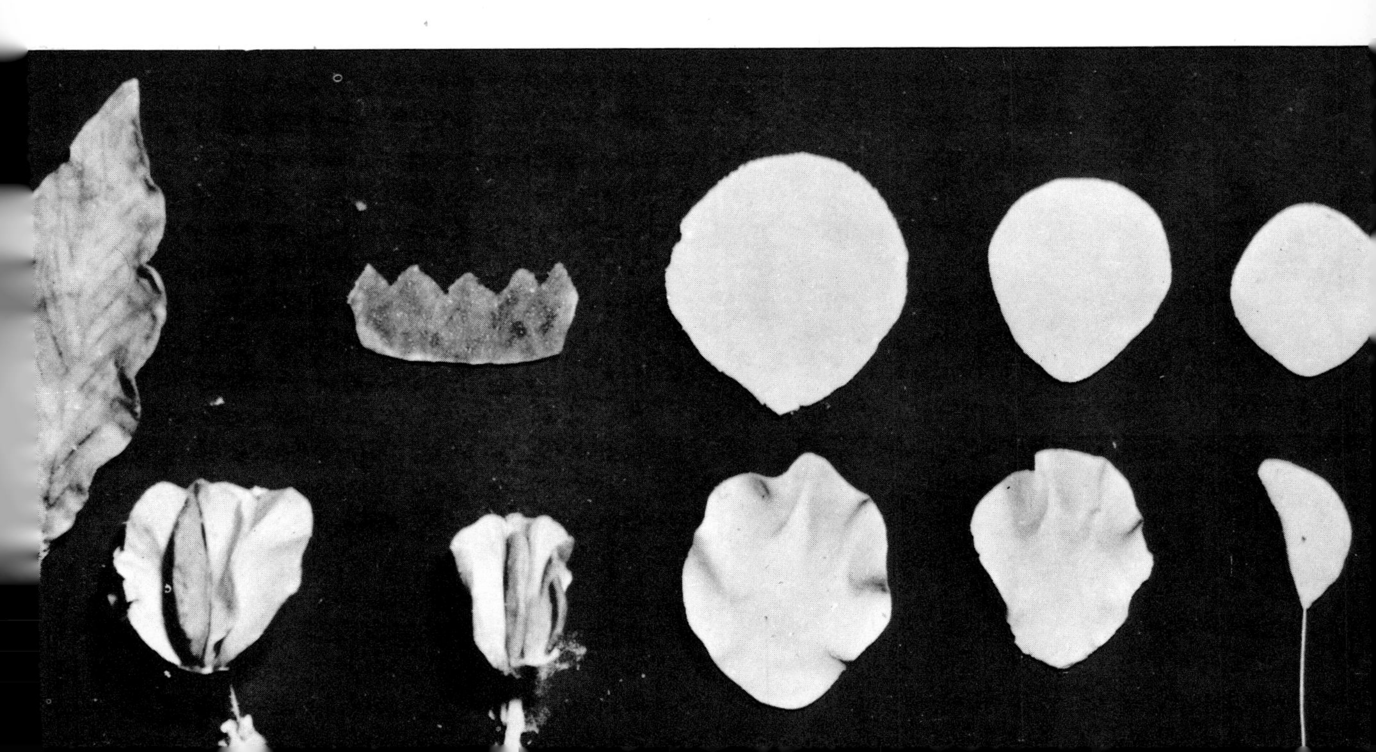

The violet and mock orange blossom step by step.

Mock Orange Blossom

This consists of four small roundish petals assembled with royal icing in a calyx in the same manner as described for the violet. A cluster of yellow stamens is added to the centre for effect see photograph below.

Moulded orange blossom in stages and assembled. This attractive little flower has five small petals, and a small fringed piece for the centre with a few stamens, which are attached to a calyx and wired.

Flannel Flower

Roll out a small piece of thin modelling paste and cut it into an oblong strip about 1½ ins. x 1 in. Then slit it to form ten petals joined at the base (see illustration page 75) and mitre the

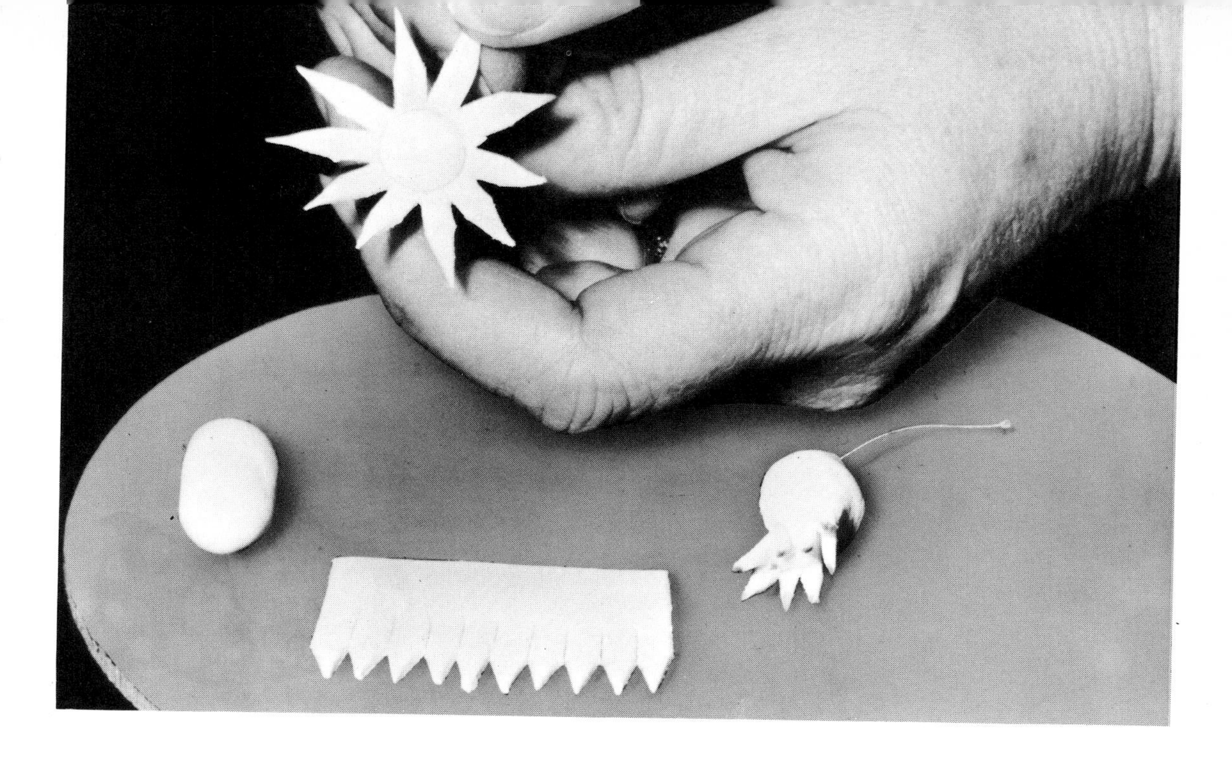

Shaping and cutting the flannel flower and its bud. Note the capsule sized piece of paste.

point of each. Use a sharp pointed knife for cutting and mark a vein in each petal. With pale green paste, roll out a long, solid cylinder approximately $1\frac{1}{2}$ ins. in circumference (so as to fit the length of the petal strip—see photograph). Dampen the uncut end of the petal section with water and wrap it around the cylinder so that both edges just meet. Turn back each petal with the fingers, pinch off any surplus underneath and insert wire if needed.

Allow the flower to dry and tip each petal point with pale green colouring. Pipe a cluster of small dots in green royal icing over the centre. A bud is obtained by leaving some petals closed over the centre section and slightly turning a couple back.

Arum Lily

The centre stamen or cone is rolled from yellow modelling paste. Then cut a large leaf shaped piece from white paste which has

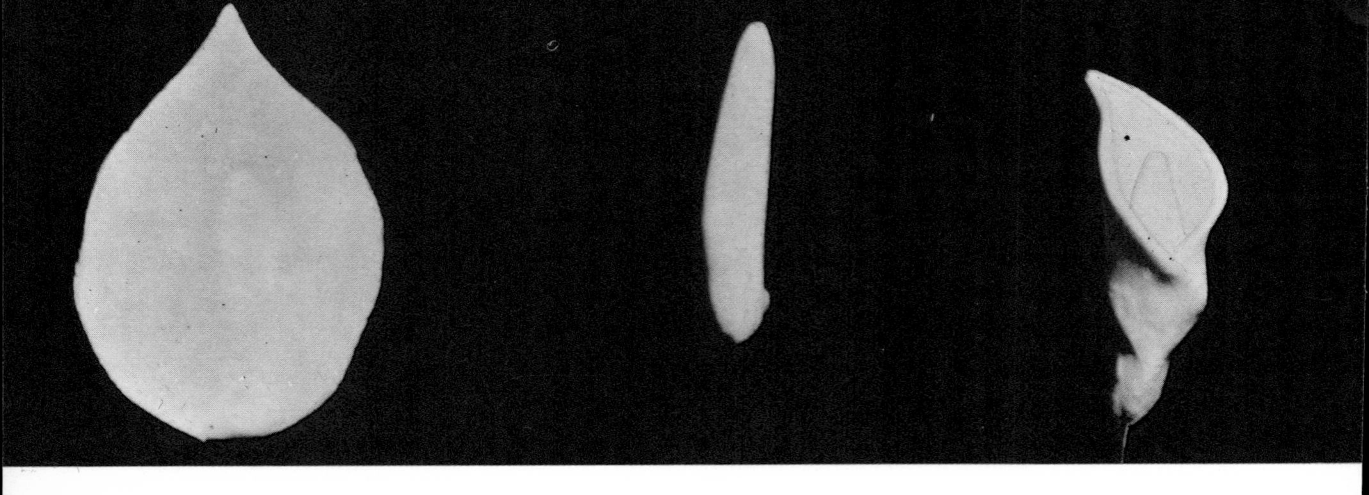

been thinly rolled, and dampen the base of this with a little water. Wrap this around the yellow centre. Pinch off any surplus and insert wire if required.

Fuchsia

Colour the modelling paste to the required shades and roll a small piece to stand as a cone. Finger out finely four small petals which have been shaped over the cushion of the thumb. Attach these in an overlapping fashion to the base of the cone with a little water. The four lower petals are cut to shape (illustrated page 77) and attached to the bell of the flower. Each of these is turned back slightly. Pinch off surplus base and add some stamens to the centre of the flower.

Water Lily

For the outer petals you need to cut seven *or* nine elongated petals approximately 1½ ins. in length and about ¾ in. in width. Place these to dry in the base of a patty tin or drape over the curve of a rolling pin.

2nd Row, cut 7 *or* 9 petals respectively again, making them smaller in length and breadth than the previous row. Allow to dry as above.

3rd Row, consists of 5 *or* 7 petals respectively, cut smaller again than the previous row; dry these over a small dowel rod or the handle of a wooden spoon. All petals must be thoroughly dry before you commence to assemble. To attach the petals you require a round base patty tin, waxed paper, No. 3 or 4 writing tube and flower stamens. Now place a small piece of waxed paper in the base of the patty tin and squeeze enough firm royal from the bag to cover a five cent piece. Position the large outer petals into this, commencing from the outside and working in towards the centre with each additional row. To attach the second row, squeeze a little more royal and place these alternately with the first, then repeat the process again for the third row, standing these in slightly more upright position. Cut the stamens in $\frac{1}{2}$ in. lengths and cluster in the centre with the help of a pair of tweezers. Allow to dry at least 12 hours before removing from the patty tin. See photographs, page 64.

The assembling of the fuschia requires delicate handling.

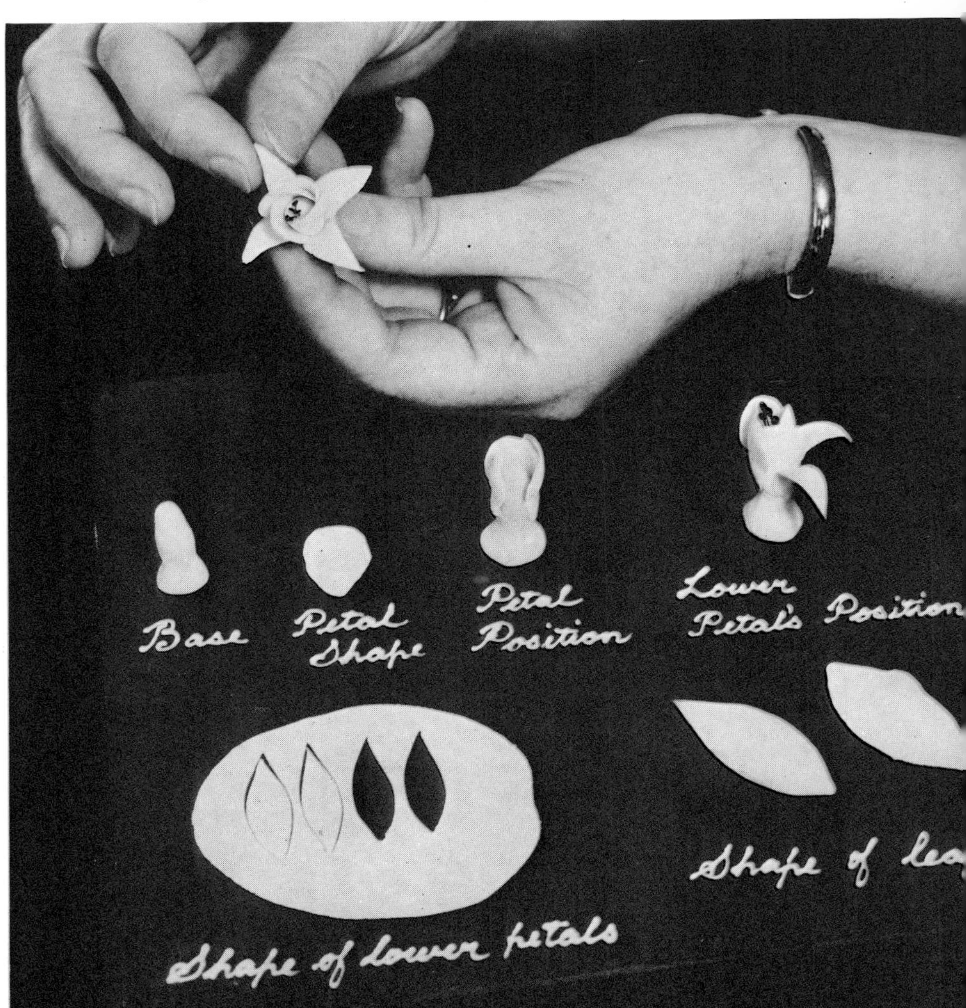

Special Occasion Cakes

THE REASONS FOR NEEDING Special Occasion Cakes can be almost without limit; we have covered the most popular, these being 21st Birthday, Children's Birthday, Mother's Day, Christening, Debutante, Anniversary, Bridal Shower, and Bon Voyage.

We feel that, should a cake be needed to suit any other occasion, the ideas obtained from the many following samples,

A simple, easy cake using the lovely Christmas bell as the main attraction. Interlaced lattice and crimper work finish the design.

coupled with the decorator's own ingenuity, would be sufficient to meet any requirements. Many feature a good assortment of hand moulded flowers, whilst others take in a variety of designs suitable for flood work. We have also taken a birthday cake from page 22 and changed it to suit a Christening cake as shown on page 110 by omitting the extension work and adding a tulle overskirt. The floral decoration has also been changed from the moulded rose to the moulded water lily. Our main purpose in so doing, is to show how our designs may be interchanged to cope with all occasions. Care should always be taken when placing the design on to the cake, as beautiful work may be marred by indifference to proper balance.

The Christmas rose and red ribbon make this a bright picture on the festive table. The rose and leaves are moulded and added to this are sprays of forget-me-nots and green fern. The base consists of scallops built out three deep, finished off with loops.

Below *The poinsettia design is a bright favourite for the Christmas cake. Green tulle holly leaves have been used in this design to give a soft effect to the moulded flower. Crimper work is used across the top and sides with a looped trimming of royal icing, and red ribbon. About 14 moulded leaves are required for poinsettia. Place into position on cake with yellow and green dots grouped in a cluster in the centre of the flower.*

Right *Here is a striking wedding cake simple enough for a beginner to decorate. The top posy consists of the elementary moulded rose, in medium and small sizes, and sprigs of lily-of-the-valley and tulle. Cut the pointed-shaped edge pattern on greaseproof paper and place it over the cake, pin-pricking the outline. Remove the pattern and pipe a small scallop above the pin line with the 00 tube. Scallop around the base about ½ in. from the level of the board with No. 1 tube. Pipe a large star beneath each, using No. 8 shell and fill in between large stars with small ones from the No. 5 tube. Above this border, for a space of approximately one inch and following the same*

scallop line, is a border of small dots and embroidery. An embroidered four-petalled flower continues in diagonal lines across top of cake. Lace edging is attached to the pattern line, and a ribbon band finishes each tier.

Here is old St. Nick at his merriest. This Christmas cake provides another example of flood work. A bright red ribbon draped across the four corners, with a touch of green and red embroidery, gives a gay finish. The design for the cake shown in this photograph is taken from a Christmas card and traced on to greaseproof paper, then transferred to the cake's surface by pin-pricking. The whisker effect is obtained by firstly flooding the area and then piping in the whiskers with royal of lattice consistency.

Good cheer seems to ring out from the sugar bells, topped with silver glitter, which adorn this Christmas cake. Moulded Christmas bells, Christmas bush, and wattle give a bright, cheery look to the spray and touches of piped flower sprigs and lace pieces complete the design.

Left *This wedding cake has maintained a dainty appearance with soft embroidery and moulded mock orange blossom. The base is something different with the heart motifs surrounded by lace, and fine extension making an attractive border. Narrow bands of ribbon cross each corner and finish with a tiny bow, adding a lift to this delicate cake. (See detail photograph on page 88.)*

Below *This wedding cake will be greatly admired by the girl who likes the look of tailored elegance in this clean cut design. Fine extension and lace is used for the base work and edged above by a band of small dots. Roses, bouvardia, tulle leaves and forget-me-nots form the floral arrangement for the corners of the cakes and top posy. Eight small hearts piped in lattice and surrounded by lace, with the bluebirds make a dainty finish.*

This Peace and Goodwill cake is suggested for Christmas whilst, with a variation in the inscription, the same design may be used at Easter or on the occasion of the First Holy Communion. A flooded church window and a spray of moulded lillies is the only decoration necessary for the top. The side design has been made more elaborate, with extension work topped by small holly sprigs. Lace pieces form a scalloped design around sides.

Another gay little cake for the festive season. Here, the design is taken from a Christmas card, and is flooded directly on to the cake. Moulded Christmas bells and flannel flowers form the spray whilst soft, piped sprigs of holly are appropriate.

Close-up of the heart motifs on base of three tier wedding cake on page 84.

A close-up view of corner floral arrangement and side pipework on the wedding cake on page 85.

This three tiered American-style wedding cake was another first prize winner at the Sydney Royal Easter Show. It featured an all over lace design with extension windows at the base. The floral arrangement contains moulded orange blossom, bouvardia, lily-of-the-valley and tulle leaves. This same arrangement forms a posy on top, softened by tulle and ribbon.

Above — *This beautiful bowl of luscious fruit is fit to grace any table. A packet of prepared marzipan is used to mould the pieces, softened by the addition of a little sweet wine so as to make it more pliable. Orange juice may replace wine to give the same effect. The fruit is moulded almost to the same size as real fruit, to add to the realism. After they have been shaped, the fruits are painted with diluted food colours. Arrange them in a natural manner over the cake which has been previously cooked in a pyrex dish, or alternatively, has been cut to a required shape.*

Right — *Designed especially for the "Official Table" this Debutante Cake would be an eye catcher. There are enough small posies on this cake to be able to present one to each debutante with her piece of cake. The posy contains roses which may be piped or moulded and small sprigs of forget-me-nots on wire. Fine lace and extension coupled with soft embroidery finish the sides.*

Left — *The main feature of this prize winning cake is the floral arrangement, consisting of one full-blown rose and smaller roses, also lily-of-the-valley and tulle leaves. Smaller sprays have been arranged on the other tiers. The embroidered design has been kept simple, to form a background. Bands of ribbon, extension work and lace complete the design.*

Below — *This heart-shaped wedding cake is a show winner. The embroidered all-over pattern is graduated from the flowers to small sprigs as the work goes up and over the sides. The spray is made up of Cecil Brunner roses, lily-of-the-valley, and tulle leaves. A dainty touch is achieved by adding tiny blue birds. Three hearts outlined with lace surround the pillars. All pipe and extension work has been done with the No. 00 tube and the built-out scallops with the No. 3.*

The tulle robe of this beautiful Christening Cake is softly piped with embroidered sprays of flowers. This is surrounded by a garland of moulded flowers containing roses, lily-of-the-valley, hyacinth and bouvardia. Fine lace and extension make a pretty base design, whilst the top is framed in an embroidered scallop featuring a ribbon insertion.

A delightful little cake for a boy or girl, with a softly embroidered pattern offset by a circle of fine lace which surrounds a pair of moulded shoes with a tulle handkerchief. The four corners have a dainty floral arrangement of roses, forget-me-nots, and tulle leaves. This also was a show winner.

This two-tier wedding cake has a 9-in. square base and a 6-in. square top. The side extension and fine lace work give a light, delicate effect. Moulded orange blossom and lily-of-the-valley have been used to form dainty sprays on the top and sides.

Basic recipe for a wedding cake

CARE SHOULD BE TAKEN to give the fullest attention to every detail concerning the making and decorating of this very special cake, so that the finished work befits the joy and admiration of this happy occasion.

A reliable recipe should be used for the baking of the cake, and we recommend the following. This is a 1¼lb. mixture.

The ingredients below are sufficient to bake a ten inch and six inch two tier wedding cake.

It is advisable to have the top tier no larger than 6 ins. otherwise the cake can look top heavy and this eliminates the dainty effect.

1¼ lbs. sultanas
1¼ lbs. currants
1¼ lbs. raisins
¾ lb. mixed peel
6 ozs. almonds
4 ozs. dates
4 ozs. glaced cherries
15 eggs
15 tablespoons brandy
1¼ lbs. butter
1¼ lbs. sugar (brown)
25 ozs. plain flour
1½ teaspoons ground ginger
1½ teaspoons mixed spice
2 tablespoons parisian essence

Method; Prepare the oven. Line tins with two or three folds of paper. Cream butter and sugar, add parisian essence. Add beaten eggs one at a time, continue beating. Add prepared fruit, and all dry ingredients. Add brandy last and mix very well. Pour into the prepared tins and bake in oven temperature 300° for first 15 minutes decreasing heat to 270° for remainder of cooking time. The large 10 in. cake takes approximately six hours to bake.

The cake should be baked five to six weeks before the

The lace design from the bride's gown was used as the embroidered pattern on this lovely three-tiered round wedding cake. The base border consists of a built-out C scroll formed with the No. 3 tube, and inside this scroll is a small piped heart. The flower arrangement is made up of frangipani and lily-of-the-valley, which is softened by tulle and ribbon.

wedding to allow time for the flavour to mature. This also ensures that the cake will cut cleanly and not crumble.

If using extension work it is an advantage to have the boards considerably larger than the cake. For instance a 10 inch cake would require a 14 inch board, the 8 inch, an 11 inch board, and the 6 inch, a 9 inch board. The board should be covered with a silver paper, or if the cake is to be used for a golden wedding celebration then gold paper is essential.

The almond paste covering should be made about three days before the plastic icing is put on, as this allows time for the oil in the almonds to dry. If this amount of time cannot be spared, dissolve a little gelatine in water and brush all over the almond paste surface. This will stop the oil from seeping through and staining the cover.

If using marzipan meal, be sure to leave it on the cake for at least 12 hours before covering with the plastic icing.

Each tier is supported by pillars, which may be purchased in plastic or gum paste according to personal choice.

Take care that each upper tier is made in proportion to the bottom tier otherwise the cake will have an unbalanced appearance. The following sizes give a good balance for a three tier cake: 10 in., 8 in. and 6 in. or 9 in., 7 in. and 5 in.

The pillars have a hollow centre through which is placed a wooden skewer. The skewer is pushed down into the cake until it reaches the board then marked and cut off to the height of the pillar. The skewers take the weight of the next tier, and save the pillars from pressing into the cake (photograph on page 99).

In this day of excellent plastic and fondant icings, it is most unwise to ever use royal icing for cake covering. This old fashioned method makes it difficult to acquire a smooth surface, and because of its hard setting manner is a nuisance to cut.

Above — *This is a similarly designed cake to the one on page 98, showing a different floral arrangement of moulded roses, hyacinths and lily-of-the-valley.*

Left — *For the bride who likes to be different this two-tier cake will make strong appeal. Three 7 in. cakes are placed on separate 8 in. boards that have been covered with silver paper, and piped with an edge of stars around the base. This is then interlaced with fine drop scallops. Moulded roses, bouvardia, lily-of-the-valley and forget-me-nots, form a most attractive floral spray.*

Moulded Dog Roses in shades of pale pink, with pink and white bouvardia, and blue forget-me-nots, in tiny sprays give a very simple design, a soft feminine look. Bluebirds add a further touch of colour.

The beautiful lace design on this wedding cake has been copied from the embroidered motif of the bride's gown. The outlined pattern is in flood work. White orchids faintly tinted with pink and lily-of-the-valley combine to make a pretty floral arrangement. All piped lace and extension work was done with 00 tube, built out base with No. 3.

Above *This beautiful wedding cake has featured the embossed rose. The large flooded rose used in conjunction with the smaller trailing rose gives a pleasing look to this cake. The floral arrangement has been kept very simple so as not to make the cake look heavy, simple sprays of bouvardia and lily-of-the-valley with small moulded roses adorning the bottom tier, and two tiny blue-birds in the centre. The large full-blown rose is arranged on top and ribbon kept to the alternate corners.*

Right *Garlands of moulded orange blossoms frame the two lower tiers of this elegant cake. Ribbon and lace have also been used lavishly to give a very dressed-up look to this beauty. All pipe work was completed with 00 tube, except base scallop of extension where No. 3 was used.*

This round American wedding cake may be baked in ten, eight and six inch tins or in nine, seven and five inch tins. Each cake is covered separately and placed on the same sized piece of waxed covered cardboard as the cake itself, and stood one on top of the other. These cakes must be kept light in design or they tend to look very heavy. This one has been sparsely scattered with an embroidered rose design piped with a fine writing tube and the same tube was used for lace edging.

An eyecatcher this three tier wedding cake. Delicately embroidered with an all over flower design, studded with tiny dots, and off-set by pastel toned pink moulded roses, accompanied by lily-of-the-valley and net leaves. A deeply scalloped base of fine lace gives a rewarding finish to the base line.

This attractive wedding cake has a 9 in. base and 6 in. top, which holds a small bridal posy containing small moulded roses, bouvardia and forget-me-nots. The top surface of both cakes is piped diagonally with an all-over embroidered flower. Small extension windows decorate the sides, with the flowers clustered at each alternate corner. Elementary moulded roses make this an easy cake for a beginner.

Small sprays of hollyhocks piped directly on to the cake give a distinctive appearance to this two-tier cake. Split ribbon has been used for side effect and the space between filled in with piped dots. A band of extension and lace work finish the base. A sheaf of moulded roses artistically grouped on the lower and top tier in pale shades of lemon complete this design.

This Christening cake was a blue ribbon first prize exhibit. The moulded water lily and all-over small flooded design are features of this lovely cake. A gathered tulle flounce held in place by a narrow ribbon band, and tiny bows, all help to give it a dainty look. We have taken this design from one of our birthday cakes to show suitability of a tulle overskirt for christening cakes.

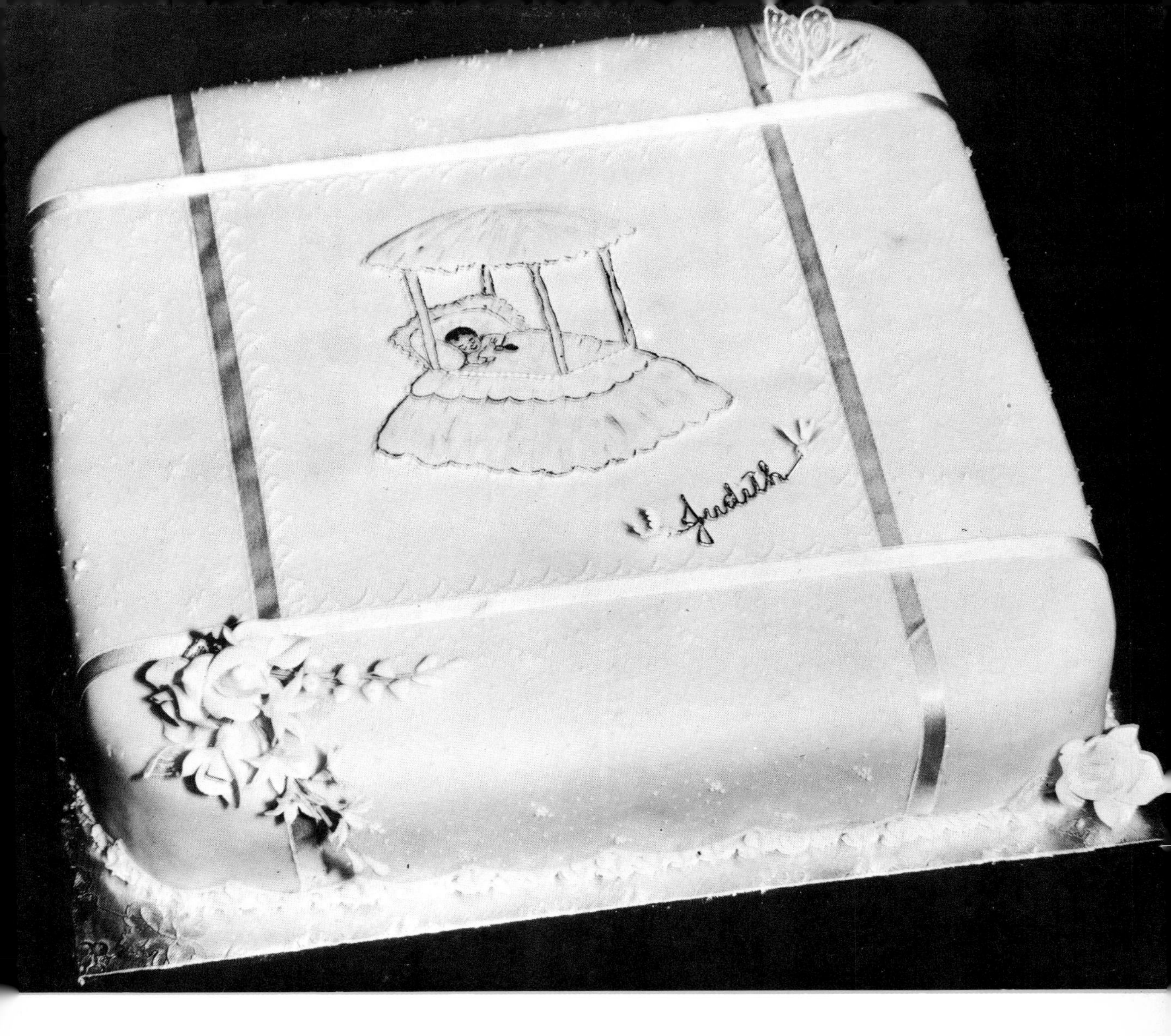

Christening cake, with scattered forget-me-nots and small dots on four sides of cake, small moulded spray of roses and hyacinths on corner and small net butterfly on top corner. The cradle may be flooded or hand painted, or any other motif may be substituted in place of this one.

A very simple little Christening cake for boy or girl. Soft embroidery over corners with cluster of Cecil Brunner roses and snow drops. The bib may be cut from fondant or flooded. An edging of lace and piped spray put finishing touches to bib.

A simple design for a quiet 21st birthday. The horseshoe is flooded and piped with fine lines and dots whilst piped apple blossom and forget-me-nots, bunched with ribbon, form floral arrangement. The name and key are painted over in gold paint. This is a most suitable cake for a man.

Lace and Embroidery Work

In the lace and embroidery field, the decorator is afforded an excellent opportunity to excel; the finer and more delicate the work, the softer and daintier the appearance. Best results will be achieved with the small bag or the writing tubes No. 0 and 00. Mastery of this work places one firmly in the professional class.

On the following pages we have taken various lace designs and in each case, have illustrated the assembly from beginning to the completed 'piece'.

The chosen design is piped directly on to the waxed paper which has been secured to a flat board with plastic tape. It will be found necessary to pipe many more 'pieces' than are required for the cake alone, so as to allow for breakages when transferring them to the cake. Removal from the waxed paper, with the aid of a knife, should be attempted only when the royal icing is thoroughly dry. A small snail-trail will secure the 'pieces' to the cake.

Both the ribbon insertion design and that of the embroidered flower (see foot of 'Lace' illustration page 116) are free-handed directly on to the cake. For ribbon insertion details see page 121.

Photograph shows different designs for decorating heart centre pieces and the lace edging outlining pattern.

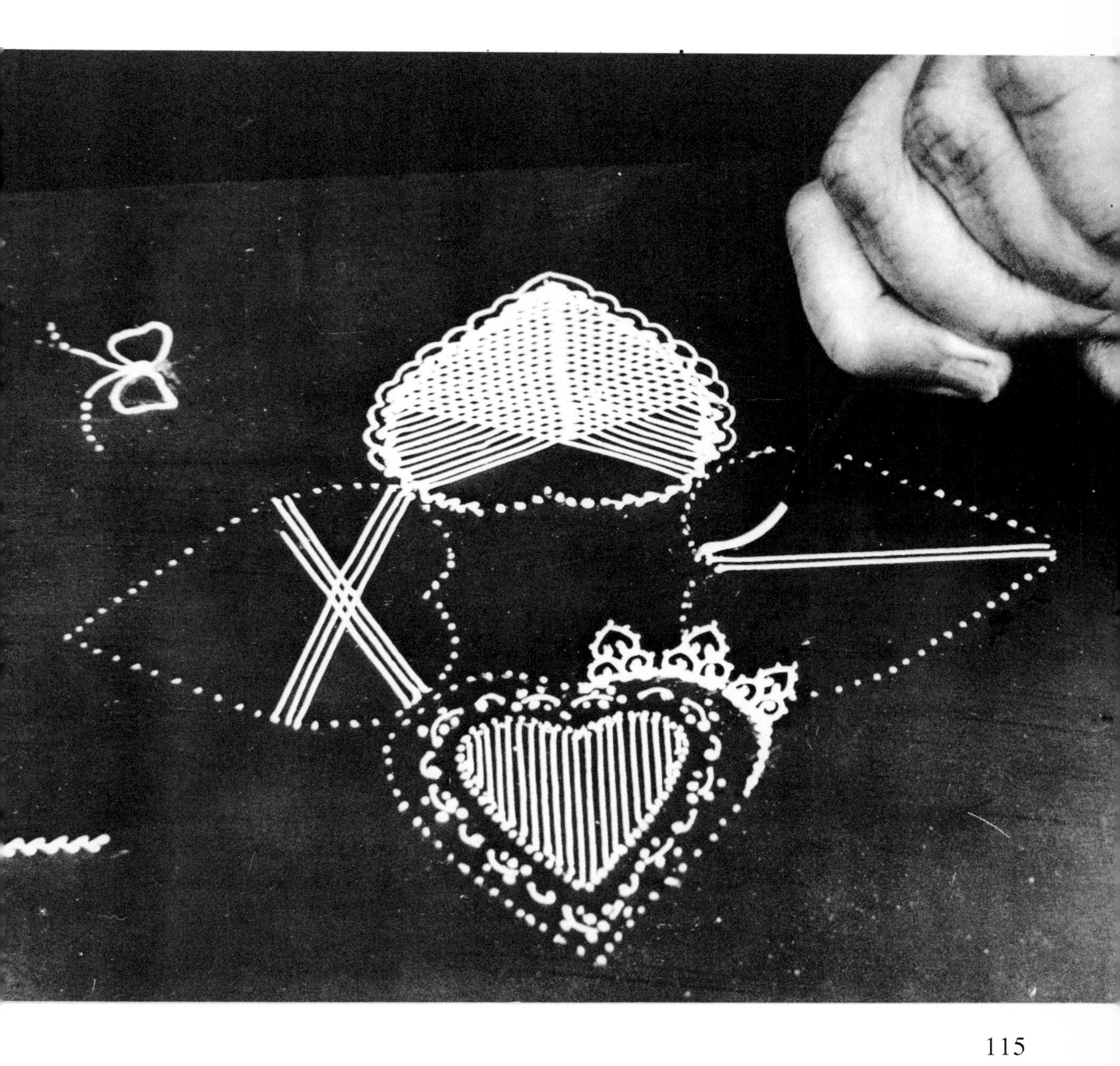

Laces

1 9

2. 10

3. 11

4 12

5 13

6 14

7 15

8 16

Embroidery

Piped Birds and Extension Work

The Assembled Bird

THIS BIRD TAKES much practice before perfection is obtained. Using the small bag, pipe an outward and back movement. The first stroke is slightly curved and the other strokes follow suit, gradually decreasing each in length and tapering at the base. These represent the feathers. Pipe the pair to correspond on waxed paper. The tail is piped with three return strokes, a long, a short and a long, without lifting the bag. The body is piped by forcing a heavy pressure flow from the bag. When the body is sufficiently long, lift the bag without breaking the flow of royal and move back slightly then forward again, releasing pressure and pulling to a sharp peak. This forms the head and beak. The dry wings and tail are then placed into position in the

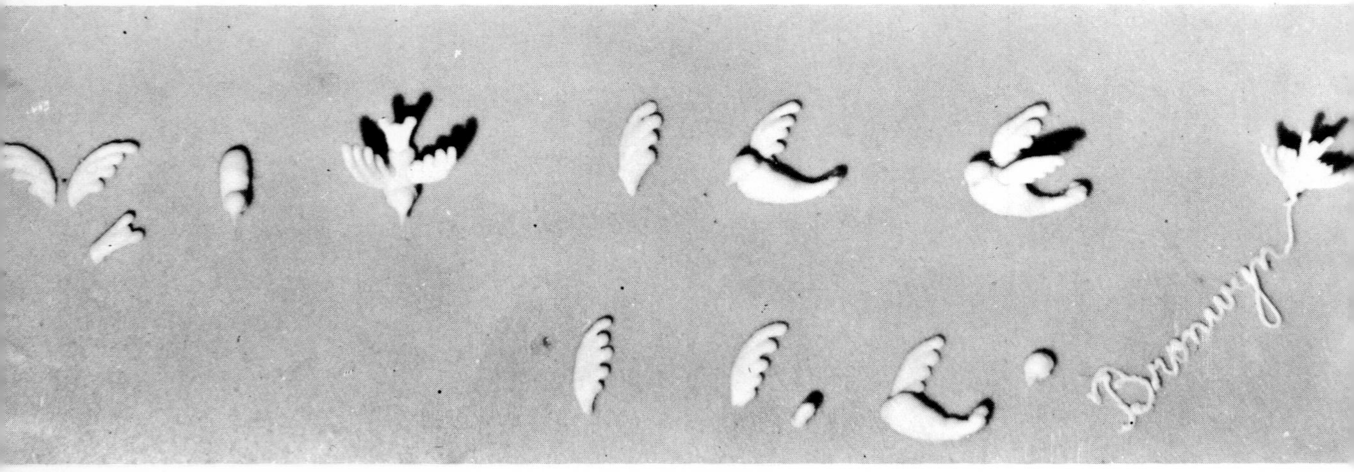

The piping and assembling of birds, used as bluebirds and doves.

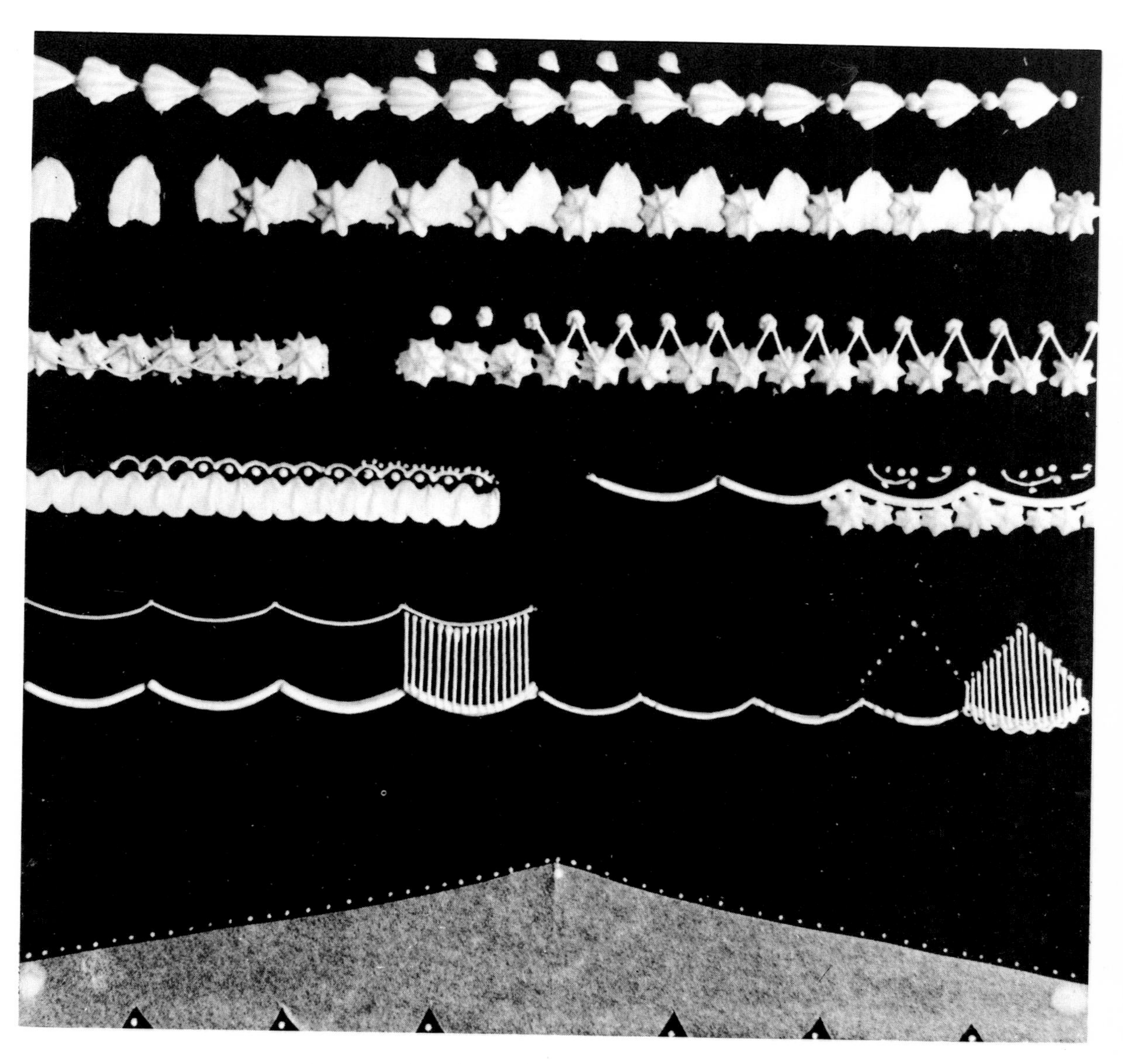

Board showing shell, star and leaf borders, star design with lacing, star with scallop and embroidered design, also small petal border with piped scallop design on top. This border is done with the 20 small petal tube. Built out extension is also shown, with the placing of the pattern for extension work on the side of the cake.

wet body. When the royal is dry the bird may be removed from the waxed paper and added to the cake.

Birds may also be drawn on paper, outlined with a soft peak consistency royal and flooded in, this is a good method for anyone who cannot master the assembled bird.

Extension and Bridge Work

This is probably the most tedious part of cake decorating.

A paper pattern may be used by folding the paper into the required amount of scallops and cutting it to shape with a pair of scissors. This is then pinned to the cake and the design pin-pricked or marked through to the plastic icing to enable the decorator to have an exact replica transferred to the cake after the paper is removed.

The lower scallop is built out from the base with a No. 3 tube and allowed to firm off well before building another piped line, directly over the first. Each row must be given time to dry or firm before the next row is piped. This process is repeated until the line has been built out five or six times. If care is not taken to proceed slowly with this operation, the decorator runs the risk of the whole work collapsing.

After this section is completed change to a 00 tube or small paper bag. Commence at the top of the pattern from the side of the cake and pipe a line to the built-out edge at the base. Continue this line slightly underneath so there will be no rough edges showing, and repeat this process all the way around cake keeping lines close together. The top of this lattice is usually neatened off with lace pieces, which have been piped directly on to waxed paper and allowed to dry, after which they are picked off with the help of a knife and placed into position on the cake with a piped line of royal. If lace is not used, the top should be neatened off with a fine snail trail or teardrop edging.

This type of work is extremely fragile, and must be handled with the utmost care.

Crimper Work

THIS EASY MEANS of trimming a cake is most popular with the advanced student as well as the beginner. Crimpers are available in different designs in small, medium or large sizes.

For use, dip the edges of the crimper into cornflour to avoid sticking; hold it at right angles to the covering, pressing together until there is a space between the prongs of approximately $\frac{1}{4}$ in. Push it into the plastic icing for a depth of approximately $\frac{1}{4}$ in. and squeeze slightly together before releasing.

For ribbon threading, make a slit with a small, sharp pointed knife to correspond with the width of the ribbon and cut the ribbon into $\frac{1}{2}$ in. lengths, folding each in half, and inserting it into the slit with the help of tweezers. To have broader ribbon, cut two slits about $\frac{1}{2}$ in. apart and insert one end of the ribbon into the first slit and the other end into the second.

For the continuous method, merely poke the ribbon into each scallop allowing a little slack; this is taken up, in each case, as the ribbon is inserted.

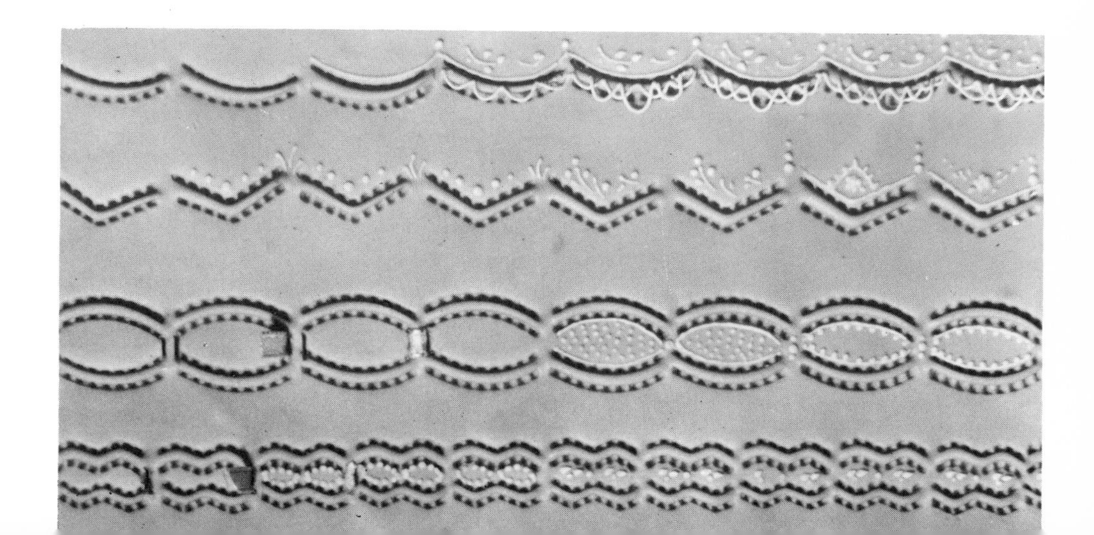

Flood Work

THIS IS A wonderful section of cake decorating for the person who possesses a good imagination and a venturesome spirit, because in this field it is possible to give full expression to your thoughts. Even the person with little imagination can produce professional likenesses to figures and designs which have been copied from books and cards. Many of the motifs in this book were copied in this manner by us.

It is advisable, upon choosing a motif, to trace it firstly on to a piece of greaseproof paper and place the tracing into the required position on the covered cake, then pin-prick the design through the paper to the cake surface. Remove the pattern and pipe these outlines directly on to the cake with '0' or '1' tube. The motif is now prepared for flooding.

Taking one colour of the design at a time, prepare sufficient royal to meet the total requirements and to this add water or lemon juice, stirring until the mixture is reduced to a thick syrup. To test the consistency, take a spoonful and allow it to pour back into the mixture. Almost instantly, the disturbed surface should level out.

With a spoon, fill a small bag to about ⅔ full (full bags become messy). Cut a small hole in it and gently squeeze out the mixture closely following the outline edge and keeping the point of the bag submerged in the royal. This helps to eliminate any air bubbles that form. Be sure to get well into the corners and crevices by using the point of the bag or by poking the royal well in with a needle or toothpick. Should a large bag be required use an 0 tube. This work must be thoroughly dry before any hand painting is attempted.

Strong, simple outline completed and flooding begun.

Pencil sketch on greaseproof paper. Outlines of design are pin-pricked through on to cake top.

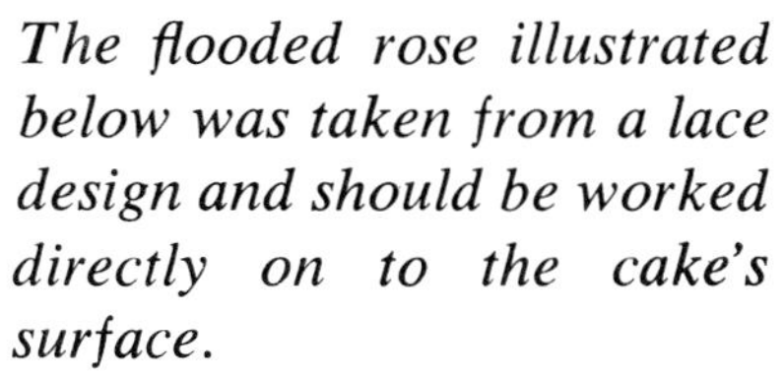

The flooded rose illustrated below was taken from a lace design and should be worked directly on to the cake's surface.

The open dog rose, always a favorite, adorns this birthday cake in beautifully arranged sprays. Two tulle bows, placed on opposite corners, add a soft effect. Many special occasion cakes such as this one, can be made using the Light Fruit Cake recipe shown on the opposite page.

Basic Recipes Hints and Show Work

Butter Cake

½ lb. butter
½ lb. sugar
¾ lb. S R flour
3 eggs
1 gill milk
vanilla essence

METHOD: Cream butter and sugar, beat eggs and add to mixture. Add milk gradually. Add sifted flour. Mix lightly and quickly. Bake in moderate oven approx. 40 minutes.

Light Fruit Cake

6 ozs. butter
6 ozs. sugar
3 eggs
10 ozs. flour (½ plain, ½ self raising)
3 tablespoons sweet sherry or milk
1 lb. mixed fruit, chopped

METHOD: Cream butter and sugar, add eggs one at a time, beat well. Add sifted flour, sherry and fruit alternately. Place in 8 in. tin and bake in moderate oven approx. one hour.

Marble Cake

6 ozs. butter
6 ozs. sugar
3 eggs
2 tablespoons milk
8 ozs. S R flour

Essence, pink colouring,
1 dessertspoon cocoa

METHOD: Cream butter and sugar, add eggs one at a time, beat well. Add sifted flour and milk alternately. Divide mixture into three parts, add a few drops of pink colouring and essence to one. Make a paste with cocoa and hot water and add to second part of mixture. Leave the remainder white.

Place dessertspoons of each mixture into an 8 in. tin, smooth surface and bake in moderate oven about 40 minutes.

Plain Cake Recipe with Cherry or Pineapple

½ lb. butter
½ lb. castor sugar
5 eggs
12 ozs. plain flour
Pinch of salt
2 level teaspoons baking powder
½ gill sherry
2 slices of crystallised pineapple or 2 ozs. of cherries
2 ozs. of chopped blanched almonds.

METHOD: Cream butter and sugar until very light in colour, add beaten eggs gradually then sherry and fruits. Stir in well the sifted flour and baking powder. Bake in 7 or 8 in. cake tin which has been lined with greased paper, and cook in a moderate oven at 350° for approximately one hour.

This recipe may be varied by using any other dried fruits, ginger or carraway seeds.

Useful Hints

Excess air bubbles through a plastic icing may be caused by the mixture being too soft. This may be remedied by the use of additional icing sugar. Another cause could be through excessive kneading, as this causes the air to become trapped in the mixture. In both cases the bubble should be pierced with a fine sewing needle and rubbed gently.

If possible, always colour by daylight, and avoid hard, bright colours—the exception being in novelty work, where the brighter shades are acceptable.

There are such high grade icing sugars on the market for present day users, that it is not necessary to add a few drops of washing blue to produce a crisper white, as the colour is usually very good. If by some chance the white cover has a dingy grey appearance, it would then be wise to add a little blue.

Easter egg moulds should never be washed, but well polished with a soft cloth; this stops the mould from sticking and reduces the chance of breaking the egg during removal.

Decorated cakes can prove a worry during wet weather and humid conditions, as they absorb the moisture from the air into the sugar and take on a moist, sweated look. This can be overcome by placing the cake in a room with a dry heat such as a gas fire, radiator, or oil heater. Don't stand the cake too close to the heat. As long as the air in the room is dry the cake will dry out also.

It is always advisable to use a fresh royal for lattice work, especially if doing extension piping.

For writing, better results will be obtained if the small paper cone is used. The point of bag or pipe should just skim the surface.

When decorating two or three tier cakes, make sure the board is covered underneath with silver or gold paper, so that no dark board is shown.

When using pillars for tiered cakes, be sure each pillar is the same length, otherwise make necessary adjustments in the height of skewers, or the cake may have a lopsided look.

Show Work

In view of the many prize winning cakes illustrated throughout this book, our work would be incomplete if we omitted to devote a few words to show that exhibiting has become one of the most popular sections among the arts and crafts at various shows. We encourage cake decorators to try their skill in this field.

By exhibiting work, it is easier to assess the value of one's artistry and workmanship in comparison with that of other top rate competitors, and to remember that constructive criticism harms no one. There is nothing like the challenge of competition to reveal weaknesses for correction and to bring out a personal style and finesse beyond expectation.

There are certain points to be considered before entering, the main one being to read the schedule thoroughly, making sure that the rules governing each section are understood.

Care must be taken to use only those materials permitted by the schedule, otherwise a possible winner could be disqualified.

It is hoped that the novice, entering a show for the first time, will be able to gain some knowledge from our experience as to the points for which a judge will be looking. We consider it important that great care should be taken with the covering which should be clean, even and smooth in texture. If a colour is to be used, it should be kept pale and delicate. On the other hand, if using white, this should be a crisp blue white appearance and not be greyish in colour. An exception to this rule is in the 'Novelty' (or, as many schedules term it 'Most Original') class where more predominant colours are acceptable depending upon the subject matter.

It is wise for the competitor to keep lace and extension work as fine as possible, and to make all other pipe work well defined and evenly executed, as the result must remain pleasing to the eye by being well proportioned and artistically presented.

Cake decorating follows fashion trends. Because of this it allows much scope for originality and it is therefore wise for the individual to develop his own personal style rather than to imitate that of someone else. By doing this a continuity of new ideas will be maintained in the art.

If using moulded flowers and leaves, care should be taken to finger them finely and to blend all colours well.

The wary exhibitor always carries some extras such as bags, tubes, royal and lace pieces, bearing in mind the Scouts' motto 'Be Prepared', and, in so doing, is ready to meet any minor emergency.

Printed by The Continental Printing Company
Limited, Hong Kong.